Biopharmaceuticals in Plants

TOWARD THE NEXT CENTURY
OF MEDICINE

Biopharmaceuticals in Plants

TOWARD THE NEXT CENTURY OF MEDICINE

KATHLEEN LAURA HEFFERON

CRC Press
Taylor & Francis Group
Boca Raton London New York

CRC Press is an imprint of the
Taylor & Francis Group, an **informa** business

CRC Press
Taylor & Francis Group
6000 Broken Sound Parkway NW, Suite 300
Boca Raton, FL 33487-2742

First issued in paperback 2017

© 2010 by Taylor and Francis Group, LLC
CRC Press is an imprint of Taylor & Francis Group, an Informa business

No claim to original U.S. Government works

ISBN-13: 978-1-4398-0474-2 (hbk)
ISBN-13: 978-1-138-11495-1 (pbk)

Library of Congress Cataloging-in-Publication Data

Hefferon, Kathleen L.
   Biopharmaceuticals in plants : toward the next century of medicine / Kathleen Laura Hefferon.
      p. ; cm.
   Includes bibliographical references and index.
   ISBN-13: 978-1-4398-0474-2 (hardcover : alk. paper)
   ISBN-10: 1-4398-0474-5 (hardcover : alk. paper)
   1. Biological products. 2. Transgenic plants. 3. Plant genetic regulation. 4.
Pharmaceutical technology. I. Title.
   [DNLM: 1. Biological Products. 2. Plants, Genetically Modified--metabolism. 3.
Gene Expression Regulation, Plant. 4. Plant Extracts. 5. Technology, Pharmaceutical.
QV 778 H461b 2010]

   QH345.H43 2010
   615'.3--dc22
                                      2009020581

Visit the Taylor & Francis Web site at
http://www.taylorandfrancis.com

and the CRC Press Web site at
http://www.crcpress.com

Contents

Preface

Plants present a novel means by which large quantities of vaccine and therapeutic proteins can be produced in a safe and cost-effective manner. Biopharmaceuticals produced in plants are easy to store, require fewer timely and expensive purification steps, and lack the containment risks associated with proteins produced in animal or bacterial expression systems. Over the past decade, much progress has been made with respect to the development of vaccines, antibodies and other therapeutic proteins. *Biopharmaceuticals in Plants: Toward the Next Century of Medicine* is a book designed to cover all major aspects of the development and production of plant-made biopharmaceuticals. The book describes the theory and practice of modern plant transformation techniques with respect to both nuclear and plastid genomes, and outlines the steps involved in the generation of transgenic plants. The engineering of plant virus expression vectors for transient expression of vaccine proteins and other therapeutics in plant tissue, and the advantages of this technology over the use of conventional transgenic plants are discussed. The significant role of glycosylation in the production of plant-made mammalian proteins and an investigation into the basis of mucosal immunity using plant-based oral vaccines is addressed. The scale-up of plant-derived vaccine and therapeutic proteins in entire crops or in large batch cell suspension cultures is covered, as is the development of clinical trials utilizing plant-derived biopharmaceutical proteins. Risks involved and biosafety concerns regarding plant-derived biopharmaceuticals are investigated in this book. *Biopharmaceuticals in Plants: Toward the Next Century of Medicine* concludes with a discussion of the future of plant-based vaccines and other therapeutic proteins in human and veterinary medicine with respect to commercial viability and as a tool to improve global public health. Do the benefits outweigh the risks? Read on, and decide for yourself.

Acknowledgments

The author would like to acknowledge the Cornell Research Foundation and the Division of Nutritional Sciences, both at Cornell University, for their assistance in the writing of this book.

The Author

Kathleen Hefferon completed her PhD in the Faculty of Medicine, University of Toronto. She worked as a post-doctoral research fellow at the University of Georgia and the Boyce Thompson Institute for Plant Research at Cornell University. She became cofounder of two start-up biotechnology companies using patents based upon her own and other research performed at Cornell University. Dr. Hefferon most recently held the title of Director of Operations, Human Metabolic Research Unit, Division of Nutritional Sciences at Cornell University. She also acts as a science writer for the Center for Hepatitis C Research at Rockefeller University in New York City and as an expert selector for the Infection and Immunity Division of the Medical Research Council, in London. She currently lives in the Fingerlakes region of New York with her husband and two children.

1 History of Plants in Production of Biopharmaceuticals

1.1 INTRODUCTION

There is virtually no place on earth where the term *transgenic plant* is unfamiliar. Transgenic plants were first developed and introduced as crops in the early 1980s. Their enormous commercial potential and role in crop improvement was fully realized upon the discovery that transformed plants were fertile and the foreign gene of interest could be continued throughout their progeny. The first genetically modified crops were soybean and corn, and appeared in the U.S. market in 1996. Since then, transgenic plants have been commercialized in many other countries. These, which exhibit increased pest and disease resistance, have been demonstrated to prevent substantial global production losses. Transgenic plants also present enormous potential to become one of the most cost-effective and safe systems for large-scale production of proteins for industrial, pharmaceutical, veterinary, and agricultural uses. In order to be effective, a plant-derived protein must be biologically identical to its native counterpart and be produced at levels high enough to be purified by relatively simple procedures.

The fact that vaccine proteins and other biopharmaceuticals can be generated rapidly and inexpensively in plants without concern for biological contamination as in animal cell culture systems presents enormous implications for the field of medicine in general. Plant-derived proteins have an advantage over protein produced in bacteria due to their ability to undergo posttranslational modifications. The possibility of producing proteins in plants offers a great opportunity to those in developing countries, where high cost, poor medical infrastructure, and lack of needles or refrigeration make vaccines difficult to come by. In this first chapter of *Biopharmaceuticals in Plants: Toward a New Century of Medicine*, the history of the first vaccine and development of vaccines in plants is described. A discussion of the rationale behind the use of plants to produce vaccine and therapeutic proteins is included. Various plant transformation techniques, with a particular focus on *Agrobacterium*-mediated

1

transformation and the regeneration of transgenic plants from cell culture is described in detail. Issues concerning the transformation of different plant species are presented. Methods used to regulate transgene expression with respect to promoter elements and the need for "plant-friendly" genes are addressed. The chapter concludes with a description of gene silencing and its effect on transgene expression.

1.2   HISTORY OF VACCINE DEVELOPMENT

By definition, the term *vaccine* refers to "any preparation used as a preventive inoculation to confer immunity against a specific disease, usually employing an innocuous form of the disease agent, as killed or weakened bacteria or viruses, to stimulate antibody production." (*Random House Unabridged Dictionary*, 1987). A vaccine against smallpox virus is often viewed as the first publicly administered vaccine and has been attributed to Edward Jenner during the late 1700s. The manner in which Jenner came to his conclusions could not have been more timely. By the end of the 18th century, smallpox was established as a major killer; mortality of those infected with smallpox was as great as 20%. Among children, 40% of deaths under 10 years of age were due to smallpox. Many of those who recovered from smallpox bore facial scars. Milkmaids, however, were often found to be resistant to smallpox and retained fair complexions. Several individuals who contracted cowpox were discovered to later not become infected upon subsequent exposure to the smallpox virus. The protective effect of cowpox was demonstrated through the inoculation of material taken from a cowpox pustule into the skin of an uninfected individual. This led to the observation that a small amount of pustule or scab from a smallpox patient resulted in a much milder infection of the disease and a correspondingly lower death rate (1:3000–1:8000 per inoculation; Gross and Sepkowitz, 1998; Baxby, 1999).

Jenner used pustular material from a cowpox lesion as an inoculum instead of similar material from a smallpox lesion. For this experiment, he inoculated cowpox lesion material from the hand of a milkmaid into the skin of an 8-year-old boy (Figure 1.1). The boy resisted variolation when challenged with smallpox about 6 weeks later. During another outbreak, several more inoculations were performed, and subsequent inoculations took place by arm-to-arm transfer of infectious material. Two years later, Jenner wrote "The Inquiry (An inquiry into the causes and effects of the variolae vaccinae, a disease discovered in some of the Western countries of England, particularly Gloucestershire, and known by the name of Cow Pox)." By 1799, Jenner's observations had been confirmed by several other practitioners, and over 1000 people had received the cowpox vaccine. Within another 3 years, the practice of cowpox inoculations had spread across Europe to North America and Asia, utilizing

FIGURE 1.1 Illustration of Jenner administering smallpox vaccine.

inoculum material provided initially from England. In spite of this rapid and phenomenal success, nearly 90 additional years were to elapse before Pasteur performed his first experiments with a potential vaccine for rabies (Henderson, 1997). Jenner referred to the infectious material as vaccine, from the Latin *vacca*, meaning cow. Many years later, Louis Pasteur in 1881 determined that all inoculations to protect against a disease be called vaccination, in honor of Jenner, and named his new discovery rabies "vaccine" despite the fact that the cow had nothing whatsoever to do with the preparation of the rabies vaccine. These experiments by Jenner, Pasteur, and many others led to what is commonly used in the vaccination process today (Baxby, 1999).

1.3 HISTORY OF VACCINE PROTEINS PRODUCED IN PLANTS

In the poorer countries of the world, where infectious diseases remain the primary cause of death, expense, inadequate health-care infrastructure, and lack of refrigeration limit the utility of vaccines. In these locations, entry of virtually all of these infectious diseases occurs through the host's mucosal surfaces in the gut, and respiratory and reproductive tracts. In 1992, an assembly of philanthropic organizations, in conjunction with the World Health Organization, set about the task of establishing the Children's Vaccine Initiative, whose focus is to advance the development of new technologies that will make novel oral vaccines accessible and advance immunization programs on a global scale (Giddings et al., 2000). Ideally, such vaccines would be cost effective, safe, and easy to store and transport. It was hoped that these new technologies would result in vaccines for diseases that were difficult to manage. Emerging technologies could also improve existing vaccination strategies by reducing the cost, removing the use of needles during immunization, and providing specific technologies for heat-stable, oral, multicomponent vaccines that required reduced or one-time administration. Among these technologies, the concept of using transgenic plants as a delivery system for edible vaccines was first developed. It has been shown that plants are capable of producing recombinant viral and bacterial antigens that undergo the appropriate posttranslational modifications to retain the same biological activity and ability to fold into quaternary structures as their mammalian-derived counterparts. Best of all, transgenic plants producing nonreplicative subunit vaccines offer an alternative that combines safety and effectiveness by enabling oral delivery through consumption of edible plant tissue (Hefferon, 2007).

In 1990, the first plant-made vaccines were performed via expression of *Streptococcus mutans* surface protein A in transgenic tobacco, followed by oral immunization of mice with the same plant material (Fischer and Emans, 2000). This transgenic plant material was later shown to successfully induce an antibody response through a demonstration that serum from immunized mice could react with intact *S. mutans*. Plants were also developed that expressed *Escherichia coli* enterotoxin B subunit (LT-B) and that exhibited successful inducement of both mucosal and serum antibody responses (Tacket, 2005). These initial experiments led to a cornucopia of studies involving generations of plant-made vaccines and therapeutic proteins and their applications in medicine.

When plant-made pharmaceuticals were first described in the general media and scientific literature, the technology was referred to as "edible vaccine." True to form, the first clinical trial performed in the United States required volunteers to consume 100–150 g of raw potato (Richter et al., 2000). Since this initial trial, researchers have speculated that plant-made pharmaceuticals can be produced in the field and consumed as a routine/local food source. In the world's developing countries, vaccines could therefore be derived from fresh produce or even in an individual's own garden. The advantages of the use of food crops for vaccine production frequently led to public misperceptions as to how these materials would be delivered in a practical sense. Eventually, to control the level of exposure of the antigen or vaccine protein, the production of plant-made vaccines and therapeutic proteins further evolved to meet the standard requirements for the production of pharmaceuticals in general by avoidance of the issues of dose variability and assurance of the high quality of the product. Edible vaccines are therefore more commonly referred to at present as plant-made vaccines, where a plant product is derived from batch-processed plant tissues or a similar processing method that can then be prescribed by a health-care worker. The final product may in fact be packaged as a capsule or juice or paste, for example, rather than as an entire fruit or vegetable. Plant-made vaccines are therefore required to comply with the regulations set out by the U.S. Food and Drug Administration and the U.S. Department of Agriculture; these regulations are still evolving within the national regulatory authorities (see Chapter 8).

Since the first vaccine protein was developed in plants, antigens from viral and bacterial mycoplasma and other pathogens, as well as antibodies specific to a wide variety of pathogens, cancers, etc., have been developed in plant tissue. In addition to this, many different therapeutic agents such as α-trichosanthin have been produced in plants. Studies have been conducted to increase protein expression levels, increase stability levels, and facilitate harvest. Synthetic genes have been constructed, and the expression of vaccine transgenes has been targeted to specific tissues. To avoid problems such as variability in protein expression levels, batch processing has been used, and the efficacy of the immune response to plant-made vaccines has been examined with respect to both the location in the body and the duration of the immune response. Both oral and nasal vaccination of plant-derived vaccines have been shown to be capable of inducing mucosal and systemic responses, and the feature of passive immunity has been demonstrated to be carried on to further offspring. Plant-made vaccines have been proven to be effective by animal antigenicity and challenge trials, and a number of human clinical trials have now been conducted. With this brief history of plant-made vaccine production to the present, it is time to turn to the subject of how plants are "transformed," that is, how a gene encoding a vaccine or therapeutic protein is incorporated into the genome of the plant.

1.4 TRANSFORMATION OF PLANT TISSUE

There are several means by which to transform plants; the most frequently used are described here. Plant transformation, meaning the stable integration of the gene of interest into a plant genome, was originally conducted using a modified strain of *Agrobacterium tumefaciens,* the bacterial strain responsible for crown-gall disease. Biolistic delivery using a device known as a "gene gun" as well as other less commonly used techniques will also be described in detail in the next section.

1.5   *AGROBACTERIUM TUMEFACIENS-* MEDIATED PLANT TRANSFORMATION

1.5.1   HISTORY AND CONCEPT OF *AGROBACTERIUM TUMEFACIENS* FOR PLANT TRANSFORMATION

The potential of *Agrobacterium*-mediated transformation to act as a conduit for foreign gene expression in plants was first considered as a possibility by Armin Braun, who in the 1949s proposed the concept of the "tumor-inducing principle" behind the mechanism of *Agrobacterium*-mediated transformation (Gelvin, 2003). Later on, in the 1970s, large plasmids were identified within several virulent strains of *Agrobacteria*. This discovery led to the recognition of a particular class of plasmids known as the "tumor-inducing" or Ti plasmids. The Ti plasmids were responsible for tumorigenesis, during which a portion of these plasmids known as the T-DNA became transferred to plant cells and incorporated into the plant genome (Gelvin, 2003).

Based on this work, the potential to use Ti plasmids as vectors to introduce foreign genes into plant cells was examined. This was accomplished using two approaches; one involved the cloning of the gene of interest into the Ti plasmid itself in such a way that the new gene was positioned in *cis* to the *vir* (virulence) genes located within the same plasmid. The other approach involved the cloning of the foreign gene into the T-region on a second, additional replicon (helper plasmid). This helper plasmid, known as a T-DNA (transfer DNA) binary vector, contained the *vir* genes as well as a deletion-mutant version of the T-region so that *A. tumefaciens* cells transformed with this plasmid had lost their ability to incite tumors (Tzfira and Citovsky, 2006).

One method for cloning foreign DNA into a Ti plasmid involves placing the gene of interest as well as an antibiotic resistance marker gene into a broad host range plasmid that possesses the ability to replicate efficiently within both *E. coli* and *A. tumefaciens* cells (Tzfira and Citovsky, 2006; Gelvin, 2003). In this protocol, all of the cloning steps take place within *E. coli* cells, and the recombinant plasmid that results is then later transformed into *Agrobacterium*. *Agrobacterium* cells that have been successfully transformed are then selected for on antibiotic-resistant agar plates. In cases where the binary-vector system is being utilized, the *vir* genes are provided on a "helper" plasmid separate from the one that harbors the T-DNA. By transforming *Agrobacterium* with both plasmids, the *vir* genes can be expressed and can act *in trans* on the T-DNA to enable transformation to take place within the plant cell.

The development of T-DNA binary vectors revolutionized the use of *Agrobacteria* to introduce genes into plants. A large number of binary vectors have been designed for highly specialized purposes and are commonly used in plant biotechnology today.

1.5.2 MOLECULAR BASIS OF AGROBACTERIUM-MEDIATED TRANSFORMATION

To understand the use of *Agrobacteria* as a tool in plant biotechnology, a thorough comprehension of the biology of this microorganism must be set in place. *Agrobacterium* as a genus is capable of transferring DNA to a diverse group of dicotyledonous and monocotyledonous plant species. *Agrobacterium* also has the capacity to transform other organisms, including yeast, fungi, and even human cells (Lacroix et al., 2006a). During infection, *A. tumefaciens* causes a mass of mainly undifferentiated cells to form on a plant's stem at the soil line (known as the crown). The T-DNA portion of the Ti plasmid and its delimiting right and left border sequences become integrated into the nuclear genome of a susceptible plant cell that is in contact with the bacterium. The T-DNA encodes enzymes for synthesizing plant hormones that stimulate cell division and the proliferation of undifferentiated cells into a tumor (Gelvin, 2003). Today, vectors used for transformation lack the genes for hormone-synthesizing enzymes and therefore can introduce foreign DNA into a nuclear chromosome of a plant cell with minimal damage. Transformation of a plant by *A. tumefaciens*

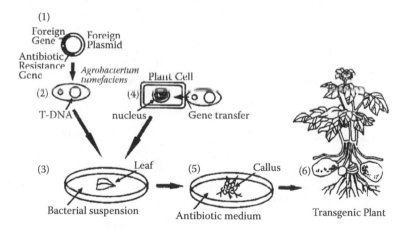

FIGURE 1.2 The stages involved in the generation of transgenic plants by *Agrobacterium*-mediated transformation. (1) The gene of interest is cloned into a foreign plasmid that contains an antibiotic resistance gene. (2) The plasmid is transformed into *Agrobacterium tumefaciens*. (3) The cut leaf is exposed to a suspension of *Agrobacteria* containing the gene of interest. (4) The gene of interest is integrated into the genomic DNA of individual leaf cells. (5) The leaf is exposed to an antibiotic to kill nontransformed cells. The surviving cells form a callus that then sprouts roots and shoots. (6) The plantlets produced from the callus are transferred to the soil. Mature transgenic plants generated now contain the foreign gene of interest.

involves the insertion of a foreign gene between the borders of the T-DNA, which in turn is cloned within a plasmid (Figure 1.2). The resulting plasmid construct is then transformed into a modified version of *A. tumefaciens* that lacks the *vir* genes. As mentioned previously, the virulence genes are supplied from the Ti plasmid itself, or else are supplied by a helper plasmid if the binary vector transformation system is used. Upon infection, the T-DNA is transferred into the plant cell, and the gene of interest is incorporated into the host chromosome. The plant cell can then be regenerated from tissue culture into a mature transgenic plant by transferal through a series of culture media with different hormone contents (Figure 1.2) (Tzifira and Citovsky, 2006).

The molecular basis of transformation is focused on the transfer of genetic material from the Ti plasmid containing the foreign gene of interest (Figure 1.3) through the nuclear membrane and the resulting integration of this genetic material into the plant chromosome. Ti plasmids range from approximately 200 to 800 kbp in size. The region of transfer,

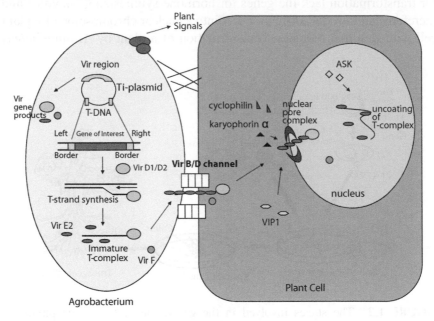

FIGURE 1.3 Model of *Agrobacterium*-mediated transformation. Various stages of transformation are depicted: (1) Attachment of *Agrobacterium* to plant cell. (2) VirD1/D2 protein complex synthesizes ssDNA known as T-strand. (3) VirE2 associates with the T-strand to form the mature T-complex that travels through a channel made of Vir proteins into the cytoplasm. (4) Host factors VIP and karyophorin α assist in transporting the T-complex through the nuclear pore complex to the nucleus. (5) Recruitment of host proteosomal degradation machinery via interaction with VirF, VIP1 and ASK1 results in uncoating and integration of the T-complex into the host genome.

known as the T-region, averages from 10 to 30 kbp in size within these plasmids. Ti plasmids can carry a single or multiple T-region. The processing of the T-DNA from the Ti plasmid and its export from the bacterium to the plant cell are the result of the activity of the *vir* genes, which are carried either within the Ti plasmid or within a helper plasmid. T-regions are delimited by their T-DNA border sequences, which are 25 bp in length, possess a high degree of homology, and flank the T-region in a directly repeated orientation (Tzifira and Citovsky, 2006; Lacroix et al., 2006a). These border sequences are the target of the *Agrobacterium* gene product VirD2, border-specific endonucleases that process the T-DNA from the Ti plasmid. Border sequences also serve as a covalent attachment site for the VirD2 protein. VirD2 is required for cleavage of the double-stranded DNA border sequences, which results from cleavage of the T-DNA "lower strand" between nucleotide 3 and 4 of the border sequence. Nicking of the border strand is associated with the covalent linkage of the VirD2 protein to the 5′ end of the resulting single-stranded T-DNA molecule known as the T-strand. This single-stranded T-strand is transferred to the plant cell, and VirD2 attaches to the right border and establishes the polarity of the molecule. The transfer of a T-strand into a plant cell and its import into the nucleus is illustrated in Figure 1.3.

1.5.3 VIR PROTEINS USED IN *AGROBACTERIUM*-MEDIATED TRANSFORMATION

The function of some of the Vir proteins produced by *Agrobacterium* during transformation has been elucidated and are described here and in Table 1.1.

1.5.3.1 VirA and VirG Proteins

VirA and VirG each function as a two-part sensory-signal transduction genetic regulatory system. VirA acts as the periplasmic antenna that senses the presence of plant-derived phenolic compounds that are induced upon wounding. It undergoes autophosphorylation and also phosphorylates the VirG protein. Upon phosphorylation, VirG activates expression of the *vir* genes, most likely by interacting with specific promoter sequences (Gelvin, 2003; LaCroix et al., 2006a).

1.5.3.2 VirD4 and VirB Proteins

VirD4 and VirB together constitute a type IV secretory system required for the transfer of T-DNA and several other Vir proteins, such as VirE2 and VirF. VirD4 may act to promote the interaction between the T-DNA/VirD2 complex with the VirB-encoded secretion apparatus. There are a total of 11

TABLE 1.1
Gene Products Involved in *Agrobacterium*-Mediated Transformation

A. *Agrobacterium*-Derived Gene Products

Gene Product	Function
VirA	Signal transduction component that senses the presence of plant-derived compounds induced upon wounding and activates VirG
VirB	Part of secretory system required for T-DNA transfer
VirD2	Part of T-DNA complex that attaches to the 5′ terminus of the T-strand and guides it through the channel into the host cell and into the nucleus; possibly involved in integration into the plant genome
VirD4	Part of secretory system required for T-DNA transfer; may also promote interaction between the T-DNA/VirD2 complex with secretory apparatus
VirE2	ssDNA-binding protein that coats the T-strand; part of the T-DNA complex; contributes to nuclear entry
VirF	Involved in T-DNA uncoating and integration into the host genome
VirG	Works in conjunction with VirA, activates vir gene expression

B. Host-Derived Proteins Involved in *Agrobacterium*-Mediated Transformation

Gene Product	Interaction	Cellular Function
Importin α (karyophorin)	VirD2	Component of nuclear import machinery
Cyclophilin	VirD2	Chaperone, assists in T-DNA trafficking
VIP1, 2	VirE2	Involved in nuclear targeting of T-complex
ASK1	VirF	Involved in mediating targeted proteolysis

VirB proteins in the secretion system; most form the membrane channel or serve as ATPases to provide energy for channel assembly or for export processes. Several of these VirB proteins form the T-pilus. The function of the T-pilus remains unclear; it may be required as the channel for T-DNA and Vir protein transfer, or it may act as a hook to grab hold of the recipient host cell and bring plant and bacterium close together for efficient gene transfer (Gelvin, 2003, LaCroix et al., 2006a).

1.5.3.3   VirD2 and VirE2 Proteins

As mentioned previously, both VirD2 and VirE2 proteins, along with the T-DNA, constitute a T-complex. It is possible that VirE2 complexes with the T-strand either within the bacterial export channel or within the plant cell. An in vitro study has suggested that VirE2 may form a pore in the plant cytoplasmic membrane to assist it with the transfer of the T-strand. VirD2, which is attached to the 5′ end of the T-strand, may then guide the T-strand through the type IV export apparatus. Since VirD2 contains a nuclear localization signal, it may direct the T-strand to the nucleus of the plant cell. In fact, VirD2 has been shown to direct DNA to the nuclei of plant, yeast, and animal cells, a result of the highly conserved mechanism of nuclear entry between diverse organisms. VirD2 also interacts with host protein karyophorin α (importin α), which is a component of the nuclear import machinery of the plant cell. Mutational analysis suggests that VirD2 also plays a role in the integration of the T-DNA into the plant genome since mutations within VirD2 can affect the efficiency or precision of T-DNA integration.

VirE2, a nonsequence-specific single-stranded DNA-binding protein, is also a nuclear protein in plant systems but fails to localize to the nucleus in yeast and mammalian cells. VirE2 may alter the T-strand from a random-coil conformation to a more elongated shape, which may assist in the movement of the T-strand through the nuclear pore. Under natural conditions, both VirD2 and VirE2 contribute to the nuclear import of T-DNA. VirD2 directs the T-DNA to the nuclear pore, and VirE2 is required for passage of the T-complex through the pore in a polar manner (Figure 1.3). The 5′ end of the T-complex carries a molecule of VirD2, whereas the rest of the T-strand is coated by VirE2. VirE2 may protect the T-strand from nucleolytic degradation. It also has been shown to interact with VIP1 (VirE2-interacting protein) in the yeast two-hybrid assay. VIP1 molecules represent host-derived molecular adapters between the T-complex and karyophorin-α-mediated nuclear import machinery of the host cell. VIP1 activity is critical for VirE2 and T-complex nuclear import. It has been suggested that VIP1 may be a limiting factor for Agrobacterium-mediated transformation of less susceptible plant species (Gelvin, 2003; Tzfira and Citovsky, 2006).

1.5.3.4 VirF Protein

Uncoating and integration into the host genome takes place after the T-complex is imported into the nucleus. One of the key Agrobacterium-derived virulence proteins involved in this process is VirF that, upon infection, is exported to the plant cell. VirF is not present in all strains of Agrobacteria, and can be considered to be a host range factor that only transforms certain host species. As an example, VirF inhibits Agrobacterium

T-DNA transfer to corn, a plant species that is difficult to transform by *Agrobacterium*.

VirF has been demonstrated to interact with ASK1, a plant homolog of the yeast Skp1-like proteins. This protein family is involved in mediating targeted proteolysis. VirF is therefore believed to target components of the T-complex to degradation by uncoating the T-strand. This targeted proteolysis occurs during *Agrobacterium* host-cell interaction. The T-strand is then converted to a double-stranded form and enters the integration pathway (Figure 1.3) (Tzfira et al., 2004).

1.5.4   OTHER GENES THAT PLAY A ROLE IN *AGROBACTERIUM*-MEDIATED TRANSFORMATION

A number of *Agrobacterium* chromosomal genes are also involved in transformation. These include genes encoding proteins involved in secretion, bacterial attachment to plant cells, exopolysaccharide production, regulation of *vir* gene induction, and T-DNA transport. A number of approaches have been used to identify plant host genes involved in the transformation process (Tzfira and Citovsky, 2002; McCullen and Binnes, 2006). One approach has been the use of the two-hybrid yeast system to identify plant proteins that interact with Vir proteins. Another involves screening for plant mutants that cannot be transformed. Host plant genes have been identified by the yeast two-hybrid system using VirD2 as bait. Through this approach, importin-α1, also known as AtKAP, a protein involved in nuclear translocation, was identified from *Arabidopsis* as playing a role in the transformation process. In another study using the yeast two-hybrid system, a VirD2-interacting protein was identified and determined to be cyclophilin. Cyclophilin is believed to serve as a chaperone to assist T-DNA trafficking within the plant cell. Further studies using VirE2 as bait uncovered VIP1 and VIP2 as interacting proteins. VIP1 and VIP2 are thought to be involved in nuclear targeting of the T-complex. Plants that overexpress VIP1 have been shown to be hypersusceptible to *Agrobacterium* transformation. Finally, VirF, a protein of unknown function, has been shown to interact with the ASK protein of the host plant. ASK may be involved in targeting proteins such as cyclins to the 26S proteosome, thus implying a role for VirF in modifying the plant cell cycle to enhance transformation (McCullen and Binnes, 2006).

1.5.5 Integration of T-DNA into the Host Genome

Little is known about T-DNA integration into the genome of the host cell; integration appears to take place in a random and nonsequence-specific manner. The number of T-DNAs and the manner in which they are inserted can be complex. For example, it is not uncommon for two T-DNA molecules to be found inserted in a head-to-head orientation within a plant genome.

It is currently believed that double-stranded break (DSB) repair enzymes may act as "baits" to attract invading T-DNA molecules to the sites of integration. This DSB repair machinery may be transported via histone modifications as part of a general process of intranuclear protein traffic that also allows transcriptional factors to reach their target promoters (Lacroix et al., 2006b).

In higher plants, the main pathway of foreign DNA integration is by illegitimate recombination (or nonhomologous recombination) (Tzfira et al., 2004; Lacroix et al., 2006b). This involves T-DNA integration at double-stranded break sites present within the host genome. A model for T-DNA integration has been determined using various *Arabidopsis* DSB repair enzyme mutants. One host cell protein, K80, has been shown to bind to double-stranded T-DNA. Complete details with respect to the mechanism of targeting T-DNA to its site of integration, the mechanism of integration, and the function of some *Agrobacterium* effector proteins remain unknown.

1.5.6 Host Range and Susceptibility of Plants to *Agrobacterium* Transformation

Genetic transformation of plants by *Agrobacterium tumefaciens* is the only known example of interkingdom genetic exchange found in nature (Gelvin, 2003). In spite of this, it is difficult to transform many agronomically important plant species using *Agrobacteria*. Plant transformation efficiency can be improved in various plant species by using more virulent strains of *Agrobacterium* or by improving plant culture conditions. Plant transformation efficiency might also be improved by manipulating the plant genome itself. For example, a screening method was developed to assist in the identification of *Agrobacterium* transformation resistant mutants (rat). One of these mutants, *rat5*, contains an insertion in the 3′ untranslated region of the *H2A-1* gene. The fact that these mutants can undergo transient expression but are unable to be stably transformed suggests that, in this case, the block to transformation lies at the step of integration. Overexpression of the *HTA-1* gene in wild-type plants was found to increase susceptibility to *Agrobacterium* infection (Gelvin, 2003; Tzfira and Citovski, 2006, 2002; Lacroix et al., 2006b).

The two-hybrid yeast system has also been used as an approach to identifying host genes involved in transformation efficiency. As mentioned previously, VirE2 has been shown to interact with VIP1 in *Arabidopsis* plants. Overexpression of VIP1 in transgenic plants significantly increased the transformation efficiency as well as the rate of transformants that can accumulate.

To date, many economically important plant species still cannot be transformed via *Agrobacterium*. The extension of host range and the requirement for an increase in transformation efficiency continues to be a major challenge (Broothaerts et al., 2005). Besides limitations in host range, there are other limitations with respect to *Agrobacterium*-mediated transformation. For example, homologous recombination would be a preferable mechanism for insertion of the gene of interest into the host chromosome, not the random T-DNA integration that takes place under the present circumstances. In addition to this, unstable transgene expression in plants remains an issue due to difficulties such as position effects and gene silencing (to be discussed later in this chapter). Further manipulation of the *Agrobacterium* genome itself should enable more efficient transformation to take place.

1.6   OTHER TRANSFORMATION TECHNIQUES

1.6.1   Biolistic Delivery

The restricted host range of *Agrobacterium* renders infection of monocots more difficult than that of dicots. Because of the limitations in host range using *Agrobacterium*-mediated transformation, other transformation procedures have been developed. The monocot maize, for example, is commonly transformed by particle bombardment, a procedure in which high-velocity microprojectiles (microcarriers consisting of subcellular-sized tungsten particles coated with the desired DNA of interest) can be "shot" with compressed helium gas using a gene gun into plant tissue (Figure 1.4). The microcarriers penetrate the plant cells, and the genes are released within. Under optimal conditions, cell injury is minimal, and the new genes are maintained within plant cells either as stable transformants or are transiently expressed (Christou, 1997; Gelvin, 2003). The gene gun was invented by John Sanford, Ed Wolf, and Nelson Allen at Cornell University.

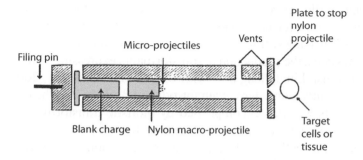

FIGURE 1.4 Diagram of a gene gun. The gene gun uses DNA-coated gold particles, precipitated on the inner wall of a plastic tube and accelerated by a flow of pressurized helium (Bio-Rad Laboratories, 1996). The most significant difference compared with other particle bombardment equipment, such as Bio-Rad's previous model, PDS-1000/He, is that the Helios™ Gene Gun requires no vacuum, removing limitations to the target and its size. Moreover, the cartridges can be stored for several months, and the bombardment procedure is much faster than with the PDS-1000/He instrument. In practice, these two particle delivery products complement each other, the vacuum chamber method providing a more controlled bombardment environment, and the Helios™ Gene Gun providing a much wider selection of target material.

1.6.2 POLYETHYLENE GLYCOL (PEG) AND OTHER TRANSFORMATION METHODS

Other techniques for transforming plants include PEG mediated trans formation and electroporation of protoplasts. The major problems in utilizing these techniques center around the difficulty in regenerating plants from single-cell protoplasts. In addition, plants that are regenerated using these techniques are often sterile and phenotypically abnormal. Other problems include fragmentation and rearrangement of genes, insertion of multiple copies of the gene of interest into the plant genome, gene duplication and rearrangement, and non-Mendelian inheritance of transgenes (Christou, 1997; Gelvin, 2003; Rakoczy-Trojanowska, 2002, Liu et al., 2006).

1.7 PROBLEMS ASSOCIATED WITH TRANSFORMATION TECHNIQUES

Foreign gene expression in nuclear transformed plants can vary markedly from one transgenic plant to another. Chromosomal position effects are partially responsible for this problem since the insertion of the transgene into the plant genome is uncontrolled. A transgene can therefore be inserted into an area of the genome consisting of chromatin that is in a transcriptionally active or inactive state. Other difficulties include the ability of nuclear transformed plants to express more than one transgene (Yin et al., 2004). Since many traits are in fact multigenic and stem from the action of several genes, the production of transformants expressing multiple genes can be a painstakingly long process (Filipecki and Malepszy, 2006).

More recently, genes have been introduced directly into the chloroplast genome. This was first accomplished by biolistic transformation. Plastid transformation is unique from nuclear transformation as the transgene is incorporated directly into the plastid genome by homologous recombination and can be predictably directed to a specific site within the plastid chromosome. Since chloroplast genes are arranged on operons, chloroplast transformation can be used to produce multicistronic mRNAs. In the future, it is possible that traits determined by multiple genes can be expressed in chloroplasts. Transgene expression levels can be several-fold higher in chloroplast-transformed plants than in their nuclear-transformed counterparts, and lack the same variation in expression levels. The sequestration of foreign proteins in chloroplasts prevents their adverse interactions with the cytoplasmic environment and protects the cell from the accumulation of potentially toxic proteins. Since chloroplasts are not present in pollen, transgenes cannot be transferred to nearby sexually compatible crops to produce "superweeds." The ability of chloroplast transformation to overcome several major problems associated with conventional nuclear technologies has created unprecedented opportunities for plant biotechnology in the future (see Chapter 3 for a detailed discussion of plastid transformation).

1.8   TRANSFORMATION OF CEREALS

Achieving success in cereal transformation has unfortunately been delayed due to difficulties that arise from the transformation of cereal plants by *Agrobacterium tumefaciens*. Specifically, cereals were originally outside the host range of *A. tumefaciens*. During the late 1980s, particle bombardment was used as the most popular method of cereal transformation. From this beginning, a variety of independent methods were developed based upon bombardment of immature embryos, and leading to the recovery of fertile transgenic plants at high frequencies (Komari et al., 1999; Christou, 1997).

As the technology for transformation evolved, it was determined that *A. tumefaciens* can indeed transform rice, maize, and other monocots efficiently if actively dividing embryonic cells (calli) are used as the transformation tissue (Sharawat and Lorz, 2006). During transformation, *Agrobacterium* must be cocultivated in the presence of acetosyringone, an inducer of *Agrobacterium* genes involved in DNA transfer. Factors involved in transformation by *Agrobacteria* include the induction of *vir* genes, choice of tissue, media, strain of *Agrobacteria*, plasmid used, and genotype of cereal (Hiei et al., 1997).

Since *Agrobacterium*-mediated transformation leads to the stable integration of a small copy number of transgenes, it is currently considered to be the method of choice. Direct transfer such as particle bombardment, on the other hand, leads to more complicated integration patterns. More recently, cotransformation of cereals using *Agrobacterium tumefaciens*-containing binary vectors that consist of two separate T-DNAs have been used with encouraging results. In this system, the first T-DNA contains a drug-resistant, selectable marker gene, and the second T-DNA contains the foreign gene of interest. Over 47% of rice transformants were generated using this superbinary vector system. Over 50% of the progeny of these transformants were free of selectable marker. This study holds great promise and advantage for cereal transformation (Hiei et al., 1997).

1.9 HAIRY ROOT TRANSFORMATION

Another transformation system that is beneficial to the production of plant-made biopharmaceuticals is the hairy root system (Guillon et al., 2006). Hairy roots can be induced from most plant species and are characterized by high branching, rapid growth rates, and the ability to develop on hormone-free medium. Transformed hairy roots can be developed as the result of the interaction between *Agrobacterium rhizogenes* and the host plant. Wounded plant tissues are inoculated with *A. rhizogenes*, which transfers T-DNA into the plant genome. The T-DNA is carried on a large root-inducing (Ri) plasmid and is carried into the host genome. Roots are inoculated with *A. rhizogenes*, and after 3 weeks, roots emerge at the wounding site, are excised, and are cultured individually on solid culture media. Each clone is then transferred to liquid culture media. Since hairy roots are genetically stable and in a differentiated state, transgenic hairy roots can be exploited almost indefinitely (Guillon et al., 2006).

1.10 SELECTABLE MARKER GENES

Selectable marker genes are used to select for plant cells that have successfully been transformed. A large number of marker genes exist; however, only a few are suitable for transformation and generation of transgenic plants. Some marker genes are not practical, and others have specific limitations, raising biosafety concerns. For example, selectable markers can pose a problem when additional foreign genes are required to be introduced into a plant genome and the number of available markers becomes limited.

Strategies have been developed to eliminate selectable marker genes. One of these strategies is to use a site-specific recombinase derived from a bacteriophage under the control of an inducible promoter in conjunction with the plant's excision machinery to excise the marker genes out of the plant genome. Alternatively, cotransformation systems can be used to eliminate marker genes. In this approach, the marker gene and gene of interest are encoded within two separate DNA molecules, and then introduced into the plant genome. The nonselectable gene can then segregate from the marker gene in the plant progeny (Miki and McHugh, 2004).

A number of selectable marker gene products are routinely used. For example, neomycin phosphotransferase [nptII] enables cells to become resistant to the antibiotics kanamycin or G418. Hygromycin phosphototransferase [hpt] enables cells to become resistant to the antibiotic hygromycin. The bar gene for phosphinotricin acetyltransferase is often used in plant transformations; this gene product confers resistance to l-phosphinotricin [PPI].

1.11   REGULATION OF TRANSGENE EXPRESSION

1.11.1   PROMOTER AND TERMINATOR ELEMENTS

All transgenes that are expressed in plants must be under some form of transcriptional control. Plants, and plant viruses, for that matter, possess their own specific promoter (DNA sequences that are recognized and bound by a DNA-dependent RNA polymerase during the initiation of transcription; highly conserved sequences within eukaryotic promoters include the TATA box) and terminator (DNA sequences that signal the termination of transcription) *cis*-acting elements that regulate endogenous gene expression. A number of these have been exploited for expression of transgenes either throughout the entire plant, at certain stages of development, or in specific plant tissues. The cloned promoters for each plant species generally retain expression profiles of their native genes both in the original as well as in other species. However, the relative activity of various promoters varies throughout plant species. A few of the most common transcriptional elements are mentioned here.

1.11.2   CONSTITUTIVE PROMOTERS

1.11.2.1   35S CaMV Promoter

The most common promoter element used in the design of transgenic plants is the 35S promoter of cauliflower mosaic virus (CaMV). This viral promoter is named for the sedimentation coefficient of its coordinating subgenomic RNA species and is constitutively expressed throughout many tissue types at high levels and in many plant species. The CaMV 35S promoter is organized in a modular fashion; a comparison of expression patterns of various combinations among *cis*-elements of this promoter have indicated that each of these elements plays a role in defining tissue-specific expression and, as a result, has pronounced implications for the combinatorial role that defines expression throughout development (Omirulleh et al., 1993; Mitsuhara et al., 1996; Benfey et al., 1990; Fang et al., 1989). Another commonly used promoter for expression of foreign genes in plants is the nopaline synthase (nos) promoter. A comparison of CaMV 35S and nopaline synthase promoters revealed that the CaMV 35S promoter was on average the stronger promoter (Sanders et al., 1987).

1.11.2.2   Ubi1 Promoter

Based on initial studies using rice transformants, it was found that the choice of promoters used to express proper foreign protein in cereals is also important. For example, the promoter element that encodes the *Ubi1* gene-encoding ubiquitin, when combined with an intron, elicited higher levels of expression of foreign protein in immature wheat, maize, and rice embryos than the 35S CaMV promoter (Christou, 1997).

1.11.3   PLANT VIRAL LEADER SEQUENCES

In order to optimize expression of proteins in transgenic plants, plant viral untranslated leader sequences are often used (Dowson et al., 1993). Plant viral leaders such as the TMV omega leader sequence can result in an enhancement of expression as great as tenfold. Mitsuhara et al. (1996) compared mRNA and protein levels and determined that mRNA containing the TMV leader sequence was translated more efficiently. This 5′ untranslated sequence of TMV is routinely used for high transgene expression (Koziel et al., 1996).

1.11.4   CHEMICAL-INDUCIBLE AND DEVELOPMENTAL PROMOTERS FOR TEMPORAL OR SPATIAL GENE EXPRESSION

In certain instances, a foreign protein may be toxic or have adverse effects on a plant with respect to its growth and physiology. In these cases, it may be advantageous to regulate expression of the foreign protein under an inducible or developmental promoter expression system. A number of chemical-inducible promoters have been isolated and used for protein production in transgenic plants. Among these are the copper-induced promoter, the tetracycline-dependent promoter, an inducible system based on rat glucocorticoid, and the ethanol-inducible promoter system (Padidam, 2003). Only the ethanol-inducible promoter has been found to be suitable for vaccine production in plants. In all other cases, the inducer either exerted phytotoxic effects, was not accessible for commercial field use, or was not sufficiently tightly regulated and therefore enabled leakage of the desired protein (Zuo and Chun, 2000).

The ethanol-inducible, or *alc*, gene expression system involves genetic elements that are derived from the filamentous fungus *Aspergillus nidulans* and that control cellular response to ethanol. This simple system involves two components, the *alc*R gene and *alc*A promoter region. The *alc*A gene encodes alcohol dehydrogenase I and is regulated by transcription factor AlcR. AlcR binds to specific sites within the *alc*A promoter region, and responds to the inducer molecule ethanol. This AlcR-mediated expression has been shown to take place in a highly responsive manner. Induction takes place within one hour and is dose dependent, with a low level of activity in the absence of the inducer (low level of leakiness), indicating that expression is under tight regulation (Roslan et al., 2001). In addition to this, direct exposure to ethanol or exposure of the plant in entirety to ethanol mist are effective means by which to induce gene expression. Maximum gene expression occurs approximately 5 days postinduction. Furthermore, ethanol, a simple organic molecule with low phytotoxicity, is biodegradable and suitable for use under commercial agricultural conditions. Besides biotechnological uses such as temporal expression of vaccine and therapeutic proteins, this system has several other applications with respect to the analysis of gene function, restoration of male fertility, and study of plant growth and development (Caddick et al., 1998; Tomsett et al., 2004).

Developmental promoters that have been commonly used for vaccine production include patatin and E8 promoters derived from potato and tomato, respectively. Patatin is one of the major soluble proteins in potato tubers and is encoded by a multigene family. The patatin promoter is tuber-specific; the protein is expressed in tubers but not in leaves (Grierson et al., 1994). Patatin has been shown to express vaccine protein in potato tubers at levels greater than the CaMV 35S promoter (Rocha-Sosa et al., 1989).

The E8 promoter is derived from the E8 gene and is expressed at high levels during tomato fruit ripening. The E8 promoter is transcriptionally activated by ethylene. Deikman et al. (1998) have shown that a fragment of 42 bp of the E8 flanking region, when fused in the forward orientation, is capable of making a minimal 35S promoter responsive to ethylene. This indicates that future transgenic plants may in fact incorporate transgenes driven by a combination of elements derived from heterogeneous promoters. To date, several vaccine proteins have been developed in tomato under the control of the E8 promoter. One recent example is cholera toxin

B protein (Jiang et al., 2007). Details of this and other examples will be discussed in Chapter 2.

1.11.5 CONSTRUCTION OF SYNTHETIC GENES

Plants, bacteria, and animal cells all contain organism-specific codon usage patterns. As a result, in certain instances, the foreign gene of interest may contain signals that are read in a plant system as cryptic termination signals or splice sites. These sequences will then be recognized by the plant transcriptional/translational machinery in a manner that differs from the organism from which the gene of interest is derived, thus resulting in a reduction of protein expression in the plant system.

To avoid improper recognition, transcription, and processing of such transcripts, synthetic genes can be designed by utilizing published codon usage tables to optimize expression in plants (Wada et al., 1992). In this way, silent nucleotide substitutions can be incorporated into the foreign sequence, and cryptic splice sites can be removed (regions containing thymine-rich intron-like sequences that could be improperly recognized by the plant cell's RNA-processing machinery) to correct the situation.

Synthetic genes can be synthesized by a technique known as oligo shuffling (Stemmer et al., 1995). A series of oligonucleotide primers can be designed that encompass the gene of interest and undergo a highly specified PCR program to produce a single product with the predicted size of the gene of interest. Synthetic genes that have been rendered more plant friendly than their native counterparts can increase processing of foreign DNA in plant cells by greater than 90% (Hefferon and Fan, 2004; Marillonnet et al., 2005). For example, a comparison of native core neutralizing epitope of porcine epidemic diarrhea virus (PEDV) indicated that the synthetic gene containing a codon usage pattern optimized for tobacco plant genes could be expressed 30-fold greater than tobacco plants expressing the native counterpart (Kang et al., 2004).

1.11.6 TARGETING OF PROTEIN TO PLANT CELL ORGANELLES

In some cases, protein accumulation and stability can be increased by the addition of a microsomal retention sequence (e.g., KDEL) to the gene of interest. This procedure is outlined in detail in Chapter 5.

1.11.7 TRANSGENIC PLANTS AND GENE SILENCING

High variation in expression levels of a particular transgene from independent transgenic plants can be due to a phenomenon in plants known as homology-dependent gene silencing (HDGS), in which transgene levels that were initially high are frequently impaired in subsequent generations. Silencing can act at both the transcriptional and posttranscriptional levels. Gene silencing involves the interaction of genes that share a homology within their promoter regions or within the transcribed regions of the silenced genes, and results in sequence-specific RNA turnover. Silencing, most likely, is the reflection of natural plant processes that control gene expression of multigene families or the interaction of plants with nucleic acids derived from invading pathogens.

Genes that have been silenced are characterized by their loss of ability to accumulate a specific RNA transcript and corresponding protein. Several silencing pathways are known to exist. Transcriptional gene silencing involves the alteration of methylation patterns of heterochromatin to a state that is inactive in the cell. Posttranscriptional gene silencing, another silencing pathway, is an intrinsic mechanism that converts double-stranded (dsRNA) into smaller (~21 nt) RNAs by cleavage with specific endonucleases called Dicers. Dicers direct the sequence-specific degradation of dsRNAs into smaller RNAs termed short interfering RNAs (siRNAs) through a process known as RNA interference. While most animals possess only one Dicer gene, plants appear to possess several; for example, four have been found in *Arabidopsis*, and six are present in rice (Waterhouse and Helliwell, 2003).

The generation of siRNAs can be induced using transgenes that express either an amplicon cassette or RNAs that contain a hairpin structure. The hairpin RNA transgene consists of a plant promoter and terminator between which an inversely repeated sequence of the target gene is inserted. RNA transcripts base pair with each other to form a hairpin structure; this dsRNA induces the silencing pathway. In addition, the importance of microRNAs (miRNAs) has recently come to the forefront as more is known about the role of gene silencing with regard to the regulation of gene expression. miRNAs are small RNAs that are encoded for primarily within the 5′ untranslated region of endogenous plant and animal genes. MicroRNAs are produced by enzymes associated within the RNAi pathway and negatively regulate several developmental processes by cleavage or translational inhibition of endogenously encoded mRNAs. Using this knowledge, synthetic mi/siRNAs can be made using the primary mi transcript and replacing the miRNA sequence with a sequence that targets the mRNA of interest. This results in the silencing of the corresponding gene. As a final note, transgenic plants may be used in the future as biofactories for production of specific siRNAs that can be applied in a clinical setting. For example, plant-derived siRNAs against influenza virus have been shown to be effective in mammalian cells (Tompkins et al., 2004).

1.12 CONCLUSIONS

This chapter dealt with the first vaccines and the history behind the concept of biopharmaceutical production in plants. Various transformation techniques were discussed with particular attention placed on *Agrobacterium*-mediated transformation. Different transformation systems, regulation of gene expression, development of synthetic genes and the effects (and potential beneficial applications) of gene silencing were presented. We now turn to an in-depth discussion of plant-made biopharmaceuticals generated via transgenic plants and virus expression systems.

REFERENCES

Baxby, D. (1999). Edward Jenner's Inquiry: a bicentenary analysis. *Vaccine* 17: 301–307.

Benfey, P.N., Ren, L., and Chea, N.H. (1990). Combinatorial and synergistic properties of CaMV 35S enhancer subdomains. *EMBO J.* 9(6): 1683–1696.

Broothaerts, W., Mitchell, H.J., Weir, B., Kalnes, S., Smith, L.M.A., Yang, W., Mayer, J.E., Roa-Rodriguez, C., and Jefferson, R.A. (2005). Gene transfer to plants by diverse species of bacteria. *Nature* 433: 629633.

Caddick, M.X., Greenland, A.J., Jepson, I., Krause, K.-P., Qu, N., Riddell, K.V., Selter, M.G., Schurch, W., Sonnewald, U., and Tomsett, A.B. (1998). An ethanol inducible gene switch for plants used to manipulate carbon metabolism. *Nat. Biotechnol.* 16: 177–180.

Christou, P. (1997). Rice transformation: bombardment. *Plant Mol. Biol.* 35: 197–203.

Deikman, J., Xu, R., Kneissl, M.L., Ciardi, J.A., Kim, K.N., and Pelah, D. (1998). Separation of cis elements responsive to ethylene, fruit development, and ripening in the 5′-flanking region of the ripening-related E8 gene. *Plant Mol. Biol.* 37(6): 1001–1011.

Dowson Day, M.J., Ashurt, J.L., Mathias, S.F., Watts, J.W., Wilson, T.M., and Dixon, R.A. (1993). Plant viral leaders influence expression of a reporter gene in tobacco. *Plant Mol. Biol.* 23(1): 97–109.

Fang, R.X., Nagy, F., Sivasubramaniam, S., and Chua, N.H. (1989). Multiple cis regulatory elements for maximal expression of the cauliflower mosaic virus 35S promoter in transgenic plants. *Plant Cell* 1(1): 141–150.

Filipecki, M. and Malepszy, S. (2006). Unintended consequences of plant transformation: a molecular insight. *J. Appl. Genet.* 47(4): 277–286.

Fischer, R. and Emans, N. (2000). Molecular farming of pharmaceutical proteins. *Transgene Res.* 9: 279–299.

Gelvin, S.B. (2003). Improving plant genetic engineering by manipulating the host. *Trends Biotechnol.* 21(3): 95–98.

Giddings, G., Allison, G., Brooks, D., and Carter, A. (2000). Transgenic plants as factories for biopharmaceuticals. *Nat. Biotechnol.* 18: 1151–1156.

Grierson, C., Du, J.S., de Torres Zabala, M., Beggs, K., Smith, C., Holdsworth, M., and Bevan, M. (1994). Separate cis sequences and trans factors direct metabolic and developmental regulation of a potato tuber storage protein gene. *Plant J.* 5(6): 815–826.

Gross, C.P. and Sepkowitz, K.A. (1998). The myth of the medical breakthrough: smallpox, vaccination and Jenner reconsidered. *Int. J. Infect. Disease* 3(1): 54–60.

Guillon, S.M., Tremoulliaux-Guiller, J., Kumar Pati, P., Rideau, M., and Gantet, P. (2006). Harnessing the potential of hairy roots: dawn of a new era. *Trends Biotechnol.* 24(9): 403–409.

Hefferon, K.L. (2007). Transgenic plants and biotechnology, in Biotechnology, [Ed. Horst W. Doelle], in *Encyclopedia of Life Support Systems* (EOLSS), Developed under the auspices of UNESCO, EOLSS Publishers, Oxford ,UK [http://www.eolss.net].

Hefferon, K.L. and Fan, Y. (2004). Expression of a vaccine protein in a cell line using a geminivirus-based replicon system. *Vaccine* 23: 404–410.

Henderson, D.A. (1997). Edward Jenner's vaccine. *Public Health Rep.* 112: 117–121.

Hiei, Y., Komari, T., and Kubo, T. (1997). Transformation of rice mediated by *Agrobacterium tumefaciens*. *Plant Mol. Biol.* 35: 205–218.

Jiang, X.L., He, Z.M., Peng, Z.O., Qi, Y., Chen, Q., and Yu, S.Y. (2007). Cholera toxin B protein in transgenic tomato fruit induces systemic immune response in mice. *Transgenic Res.* 16(2): 169–175.

Kang, T.J., Kang, K.H., Kim, J.A., Kwon, T.H., Jang, Y.S., and Yang, M.S. (2004). High-level expression of the neutralizing epitope of porcine epidemic diarrhea virus by a tobacco mosaic virus-based vector. *Protein Expr. Purif.* 38(1): 129–135.

Komari, T., Hiei, Y., Ishida, Y., Kumashiro, T., and Kubo, T. (1999). Advances in cereal gene transfer. *Curr. Opin. Plant Biol.* 1: 161–165.

Koziel, M.G., Carozzi, N.B., and Desai, N. (1996). Optimizing expression of transgenes with an emphasis on post-transcriptional events. *Plant Mol. Biol.* 32(1–2): 393–405.

Lacroix, B., Li, J., Tzfira, T., and Citovsky, V. (2006a). Will you let me use your nucleus? How Agrobacterium gets its T-DNA expressed in the host plant cell. *Can. J. Pharmacol.* 84: 333–345.

Lacroix, B., Tzfira, T., Vainstein, A., and Citovsky, V. (2006b). A case of promiscuity: Agrobacterium's endless hunt for new partners. *Trends Genet.* 22(1): 29–37.

Liu, Y., Yang, H., and Sakanishi, A. (2006). Ultrasound: mechanical gene transfer into plant cells by sonoporation. *Biotechnol. Adv.* 24(1): 1–16.

Marillonnet, S., Thoeringer, C., Kandzia, R., Klimyuk, V., and Gleba, Y. (2005). Systemic *Agrobacterium tumefaciens*-mediated transfection of viral replicons for efficient transient expression in plants. *Nat. Biotechnol.* 23(6): 718–723.

McCullen, C.A. and Binnes, A.N. (2006). *Agrobacterium tumefaciens* and plant cell interactions and activities required for interkingdom macromolecular transfer. *Annu. Rev. Cell Dev. Biol.* 22: 101–27.

Miki, B. and McHugh, S. (2004). Selectable marker genes in transgenic plants: applications, alternatives and biosafety. *J. Biotechnol.* 107(3): 193–232.

Mitsuhara, I., Ugaki, M., Hirochika, H., Ohshima, M., Murakami, T., Gotoh, Y., Katayose, Y., Nakamura, S., Honkura, R., Nishimiya, S., Ueno, K., Mochizuki, A., Tanimoto, H., Tsugawa, H., Otsuki, Y., and Ohashi, Y. (1996). Efficient promoter cassettes for enhanced expression of foreign genes in dicotyledonous and monocotyledonous plants. *Plant Cell Physiol.* 37(1): 49–59.

Omirulleh, S., Abraham, M., Golovkin, M., Stefanov, I., Karabaev, M.K., Mustardy, L., Morocz, S., and Dudits, D. (1993). Activity of a chimeric promoter with the doubled CaMV 35S enhancer element in protoplast-derived cells and transgenic plants in maize. *Plant Mol. Biol.* 21(3): 415–428.

Padidam, M. (2003). Chemically regulated gene expression in plants. *Curr. Opin. Plant Biol.* 6(2): 169–177.

Rakoczy-Trojanowska, M. (2002). Alternative methods of plant transformation: a short review. *Cell Mol. Biol. Lett.* 7(3): 849–858.

Random House Unabridged Dictionary, Copyright © 1997, by Random House, Inc., on Infoplease.

Richter, L.J., Thanavala, Y., Arntzen, C.J., and Mason, H.S. (2000). Production of hepatitis B surface antigen in transgenic plants for oral immunization. *Nat. Biotechnol.* 18(11): 1167–1171.

Rocha-Sosa, M., Sonnewald, U., Frommer, W., Stratmann, M., Schell, J., and Willmitzer, L. (1989). Both developmental and metabolic signals activate the promoter of a class I patatin gene. *EMBO J.* 8(1): 23–29.

Roslan, H.A., Salter, M.G., Wood, C.D., White, M.R., Croft, K.P., Robson, F., Coupland, G., Doonan, J., Lauifs, P., Tomsett, A.B., and Caddick, M.X. (2001). Characterization of the ethanol-inducible alc-gene expression system in *Arabidopsis thaliana. Plant J.* 28(2): 225–235.

Sanders, P.R., Winter, J.A., Barnason, A.R., Rogers, S.G., and Fraley, R.T. (1987). Comparison of cauliflower mosaic virus 35S and nopaline synthesis promoters in transgenic plants. *Nucleic Acids Res.* 15(4): 1543–1558.

Sharawat, A.K., and Lorz, H. (2006). Agrobacterium-mediated transformation of cereals: a promising approach crossing barriers. *Plant Biotechnol. J.* 4(6): 575–603.

Stemmer, W.P.C., Crameri, A., Ha, K.D., Brennan, T.M., and Heyneker, H.L. (1995). Single-step assembly of a gene and entire plasmid from large numbers of oligodeoxyribonucleotides. *Gene:* 49–53.

Tacket, C.O. (2005). Plant-derived vaccines against diarrheal diseases. *Vaccine* 23(15): 1866–1869.

Tompkins, S.M., Lo, C.Y., Tumpey, T.M., and Epstein, S.L. (2004). Protection against lethal influenza virus challenge by RNA interference in vivo. *Proc. Natl. Acad. Sci. U. S. A.* 101(23): 8682–8686. Epub June 1, 2004.

Tomsett, B., Tregova, A., Garoosi, A., and Caddick, M. (2004). Ethanol-inducible gene expression: first step towards a new green revolution? *Trends Plant Sci.* 9(4): 159–161.

Tzfira, T. and Citrovsky, V. (2002). Partners-in-infection: host proteins involved in transformation of plant cells by Agrobacterium. *Trends Cell Biol.* 12(3): 121–129.

Tzfira, T., Li, J., Lacroix, B., and Citovsky, V. (2004). Agrobacterium T-DNA integration: molecules and models. *Trends Genet.* 20(8): 375–383.

Tzfira, T., Citovsky, V. (2006). Agrobacterium-mediated genetic transformation of plants: biology and biotechnology. *Curr. Opin. Biotechnol.* 17: 147–154.

Wada, K., Wada, Y., Ishibashi, F., Gojobori, T., and Ikemura, T. (1992). Codon usage tabulated from the GenBank genetic sequence data. *Nucleic Acids Res.* 20: 2111–2118.

Waterhouse, P. and Helliwell, C. (2003). Constructs and methods for high-throughput gene silencing in plants. *Methods* 30: 289–295.

Yin, Z., Plader, W., and Malepszy, S. (2004). Transgene inheritance in plants. *J. Appl. Genet.* 45(2): 127–144.

Zuo, J., and Chun, N.H. (2000). Chemical-inducible systems for regulated expression of plant genes. *Curr. Opin. Biotechnol.* 11(2): 146–151.

2 Transgenic Plants Expressing Vaccine and Therapeutic Proteins

2.1 INTRODUCTION

In poorer countries of the world, where infectious diseases remain the primary cause of death, expense, inadequate health-care infrastructure, and lack of refrigeration limit the utility of vaccines. The entry of virtually all of these infectious diseases occurs through the host's mucosal surfaces of the gut, and respiratory and reproductive tracts. In 1992, an assembly of philanthropic organizations, in conjunction with the World Health Organization, set about the task of establishing the Children's Vaccine Initiative, whose focus is to advance the development of new technologies that will make novel oral vaccines accessible on a global scale. Ideally, such vaccines would be cost effective, safe, and easy to store and transport. Among these technologies, the concept of using transgenic plants as a delivery system for edible vaccines was first developed. It has been shown that plants are capable of producing recombinant viral and bacterial antigens that undergo the appropriate posttranslational modifications to retain the same biological activity and ability to fold into quaternary structures as their mammalian-derived counterparts. Best of all, transgenic plants producing nonreplicative subunit vaccines offer an alternative that combines safety and effectiveness by enabling oral delivery through consumption of edible plant tissue.

Recombinant secretory antibodies, stable in the harsh environment of the mucosal system, have also been expressed in plants for the purpose of passive immunization. Subunit vaccines expressed in transgenic plants have been assessed in clinical trials, and the results have been promising. Antibodies and other therapeutic agents such as interleukins are also being produced in plants. This chapter provides a cornucopia of examples of vaccine and other biopharmaceutical proteins produced in plants. Applications of this technology range from anticancer agents to contraceptives, and are discussed in detail here. The chapter ends with a description of recent biotechnological efforts made to enhance the nutritional content and/or medicinal properties of plants.

2.2 TRANSGENIC PLANTS EXPRESSING VACCINES AGAINST DIARRHEAL DISEASES

2.2.1 LT-B AND CT-B

Since diarrheal disease is an important cause of mortality among children in developing countries, preliminary research on vaccine production in plants focused on designing vaccines to protect against pathogens that cause diarrhea. The primary pathogens responsible for acute watery diarrhea are enterotoxigenic *Escherichia coli* and the related organism, *Vibrio cholerae*. Both colonize the epithelium of the small intestine and produce enterotoxins, heat-labile enterotoxin (LT) and cholera toxin (CT), respectively, which are responsible for the diarrheal symptoms. X-ray crystallography analysis indicates that both LT and CT are composed of one A subunit (27 kDa) containing the toxic ADP-ribosylation activity, and five B subunits (11.6 kDa) that self assemble into a pentameric ring structure. Specific binding of the nontoxic pentamer of LT-B to Gm_1 gangliosides present on the surface of gut epithelial cells allows entry of toxic LT-A into these cells (see also Chapter 7 on mucosal immunity).

Both LT-B and CT-B can act as oral immunogens and can induce a mucosal antibody response. However, production of B subunit vaccines in yeast or bacteria requires large-scale fermentation and costly purification protocols. When administered orally, LT-B elicits a strong oral immune response, resulting in the appearance of anti-LT-B immunoglobulins in the serum (IgG and IgA) and in mucosal secretions (secretory IgA or sIgA). Secretory antibodies in mucosal fluids prevent LT-B binding to epithelial cells, thus interfering with the toxic effect of LT.

Transgenic plants producing LT-B can produce assembled oligomers that mimic the structure of the authentic LT-B. LT-B expressed in transgenic plants has been shown to be partially pentamerized and can bind to gangliosides. LT-B expression in transgenic plants was enhanced by the addition of a plant-cell microsomal retention sequence to target LT-B to the endoplasmic reticulum. Also, replacement of the LT-B gene with a synthetic, plant-optimized gene eliminates spurious mRNA processing signals and produces an authentic amino acid sequence specified by plant-preferred codons. The ability of plant-derived rLT-B to act as an oral immunogen was tested in mice and compared with bacterial-derived rLT-B. Plant-derived rLT-B was either dispensed to mice by gastric intubation (gavage) or by oral administration of raw transgenic potato tubers. Both humoral and mucosal immune responses were induced in these mice with titers comparable to bacteria-derived rLT-B, and these antibodies were able to neutralize LT activity.

Since this study demonstrates that transgenic plants expressing a foreign antigen can induce oral immunization, a transgenic line of potatoes was selected for preclinical trials to determine the extent of protection from LT toxin or bacterial challenge. Animals fed potato tubers produced LT-B IgG in serum and gut mucosal sIgA. Mice fed with transgenic potato tubers expressing LT-B, then challenged with a gavage of LT, were shown to accumulate less fluid in the gut than control mice. This study demonstrates the feasibility of using transgenic plants as expression and delivery systems for oral vaccines and prompted the initiation of a human clinical trial, which is described in detail in Chapter 7.

In 1997, the U.S. Food and Drug Administration approved human clinical testing of raw potatoes containing LT-B. Fourteen volunteers ingested either 50 g or 100 g of transgenic potato, or 50 g of nontransformed control potato (equivalent to 0.5 or 1 mg per dose). The appearance of gut-derived antibody-secreting cells in the circulation 7–10 days after each immunization reflects priming of the gut mucosal immune system. A fourfold rise in IgG concentration was observed in the sera of 10 of 11 volunteers who ingested transgenic potato, and 6 of 11 volunteers displayed a fourfold rise in IgA anti-LT, as determined by neutralization assays. None of the volunteers who consumed control, nontransformed plants displayed elevated IgG or IgA levels. IgG levels in all respondents remained elevated when tested again at 59 days after ingestion of the first dose. LT-B-specific IgG and IgA levels were at an amplitude comparable to a challenge with 10^9 virulent enterotoxigenic *E. coli,* an amount sufficient to induce severe diarrhea. Collected stool was assayed for the presence of sIgA anti-LT. Half of all volunteers developed fourfold elevations in sIgA levels, while levels remained the same for volunteers who consumed control tubers.

In another study, Moravec et al. (2007) expressed LT-B in the endoplasmic reticulum (ER) of soybean storage parenchyma cells. Soybean is an effective vaccine expression platform for oral delivery. LT-B was found to accumulate approximately 2.4% of total seed protein and was stable in desiccated seed. Soybean extracts were orally administered to mice, and induced both systemic IgG and IgA as well as mucosal IgA responses. This plant-derived vaccine was particularly efficacious when used in a parenteral prime-oral gavage boost immunization regime. Immunized mice exhibited partial protection against LT challenge. Furthermore, transgenic soybean stimulated an antibody response against a coadministered antigen 500-fold.

Cholera is a devastating infectious diarrheal disease that affects over 5 million people and causes the deaths of 200,000 annually. As with LT-B, potato plants expressing CT-B induced significant levels of neutralizing anti-CT-B, sufficient to generate protective immunity against the biological effects of CT. Furthermore, immunoglobulin levels in orally immunized mice could be boosted by additional oral doses, implying that edible vaccines may be used to boost immune responses after primary oral or even parenteral vaccination. As with LT-B, plant-derived CT-B is less effective in stimulating an immune response than the same amount of bacterial CT-B, most likely due to differences in antigen delivery.

CT-B has also been used as a carrier molecule to promote gut mucosal immunization with other antigens. To prevent spontaneous autoimmune diabetes, oral administration of pancreatic tissue-specific autoantigens is necessary. However, the autoantigen is required in large quantities. Due to its affinity for the cell surface receptor GM_1-ganglioside located on gut epithelial cells, CT-B was employed as a carrier molecule to target insulin to the gut mucosal immune system (see Chapter 7). A plant-based cholera toxin B subunit–insulin fusion protein produced in transgenic plants was found to protect against the development of autoimmune diabetes in NOD (nonobese diabetic) mice fed with tubers. These mice displayed an induction in levels of both systemic and intestinal antibodies, while mice fed transgenic plants producing insulin alone showed no effect.

2.2.2 NORWALK VIRUS

The Norwalk virus (NV), a member of the Calciviridae family, causes epidemic acute gastroenteritis in humans. In the United States, 42% of all outbreaks of acute epidemic gastroenteritis are caused by Norwalk and Norwalk-like viruses. The virus contains 180 copies of a single capsid protein of 58 kDa, which, when expressed in transgenic tobacco plants, assemble into virus-like particles (VLPs) indistinguishable from those obtained from VLPs expressed and purified from insect cell cultures. NV CP, produced in transgenic tobacco, cross-reacted with sera derived from Norwalk virus-infected humans. Mice gavaged with NV VLPs purified from transgenic tobacco plants or fed with transgenic tubers expressing viral capsid protein elicited both mucosal and humoral antibody response against NV CP, including pronounced increases in IgA, although a better response was elicited in the mice gavaged with purified VLPs. This discrepancy may be explained by the fact that, in tubers, only 50% of NV CP were assembled into VLPs, thus decreasing the stability and immunogenicity of the presented antigen. Capsid proteins of many viruses can assemble into virus-like particles and, consequently, mimic the structure of authentic viral proteins with respect to morphology, antigenic properties, and stability. These particles are capable of eliciting both serum IgG and fecal IgA responses in mice. In a later study, NV CP was produced in tomato at 20 µg/g fruit mass. Mice that were fed with freeze-dried tomato powder expressing NVCP also stimulated both high IgG and IgA responses (X. Zhang et al., 2006). Furthermore, it was found that air-dried tomato stimulated a stronger immune response than freeze-dried fruit.

2.2.3 ROTAVIRUS

Dong et al. (2005) expressed a codon-optimized version of the VP6 protein of rotavirus in transgenic alfalfa. Gavaged female mice produced high titers of anti-VP6 serum IgG and mucosal IgA. Importantly, the offspring of immunized dams developed less severe diarrhea after challenge with simian rotavirus SA-11, demonstrating passive protection from dams to pups. The results of this study suggest that transgenic alfalfa provides a means of protecting children from severe acute rotavirus-induced diarrhea.

In 2006, J.T. Li et al. demonstrated that the VP7 gene of rotavirus transformed into the potato genome remains hereditarily stable for over 50 generations. Transgenic plants cultivated to the 50th generation that expressed this viral protein were able to elicit both humoral and mucosal responses in BALB/c mice with no significant difference in the immune response between mice fed with the first or last generation of transgenic potato.

2.2.4 HEPATITIS B VIRUS (HBV)

HBV is one of the major causes of chronic viremia and is responsible for the subsequent development of chronic liver disease. Since over 300 million people worldwide may be carriers of HBV, an effective vaccine is required to limit further HBV infection. VLPs averaging 22 nm in diameter can be recovered from subjects infected with HBV, and these have been further characterized as consisting of the surface antigen (HBsAg). The first vaccine against HBV was developed from the plasma of HBsAg-positive individuals. However, the inability of HBV to be propagated in tissue culture, as well as a number of safety concerns, imposed major barriers for the utilization of this vaccine for routine immunization. Currently, recombinant yeast-derived HBsAg (rHBsAg) is the material available for a vaccine against HBV. However, the expense of producing this vaccine in large quantities and corresponding to the immunization programs that would be necessary prohibit the use of this vaccine in the developing world. The development of transgenic plants that express HBsAg is potentially a more cost-effective way of successfully combating the disease on a global scale.

HBsAg was the first viral antigen to be produced in transgenic plants. The protein self-assembles into subviral mammalian particles of 22 nm, and is virtually indistinguishable from serum-derived and yeast-derived HBsAg in both infected sera as well as commercial vaccines with respect to size, density sedimentation, and immunogenicity.

Levels of HBsAg of up to 66 ng/mg of total soluble protein have been found in leaves of transgenic plants. To facilitate oral delivery, HBsAg has also been expressed in lettuce leaves and cherry tomatillos. The highest levels of HBsAg were found in transgenic potato at levels of 16 µg/g tuber by optimizing the codon usage pattern and including subcellular targeting signals. HBsAg has also been produced at 22 mg/L in soybean cell culture.

Since plants were shown to produce immunologically reactive HBsAg, HBsAg-specific antibodies were next shown to be generated in BALB/C mice. rHBsAg was first used as a crude extract to immunize mice parenterally. An immune response, similar to that induced by the current yeast-derived vaccine commercially available, was observed. The fidelity of the T-cell epitopes was determined by extracting T-cells from lymph nodes of mice primed by parenteral immunization with tobacco-derived rHBsAg. These cells were shown to proliferate in vitro by stimulation with both plant- and yeast-derived rHBsAg. Preservation of the epitope demonstrated that HBsAg produced in transgenic plants can mimic the commercial vaccine that is currently available. Oral immunogenicity studies in mice have been performed and are discussed in detail in Chapter 7.

In transgenic potato leaves, HBsAg was found inside membrane vesicles as 17 nm particulate structures. In potato tubers, HBsAg accumulates in tubular structures within the ER membrane. When intraperitoneally delivered, VLPs composed of HBsAg extracted from transgenic tobacco stimulated antibody and T-cell responses in mice. Feeding mice raw pieces of potato primed and boosted high-level anti-HBsAg IgG responses (in conjunction with cholera toxin adjuvant). Two human clinical trials have been conducted and are described in more detail in Chapter 7. In general, these trials have demonstrated that plants expressing HBsAg can elicit an immune response. Ingestion of transgenic potato tubers in volunteers previously vaccinated against clinical HBV resulted in significant increases in serum anti-HBsAg titers.

In addition to this, virus-like particles have been generated that contain a form of HBsAg that has been modified at the N-terminus so that it can be utilized to present T- and B-cell epitopes. The fact that the VLPs remained intact suggests that this alteration did not negatively affect the antigenic properties of the protein and that a multivalent response could be made possible.

Plants expressing the HbcAg core antigen have also been shown to self-assemble into nucleocapsid particles that contain both the viral polymerase and RNA. The particles have an immunoenhancing effect when codelivered with HBsAg, but do not alone protect against HBV infection. A deconstructed tobacco mosaic virus (TMV) vector system (see Chapter 4) has been utilized to express HBcAg in plants. In this context, levels of HBsAg were determined to be up to 7% TSP and were capable of stimulating serum

IgG responses in mice, suggesting that this VLP composed of HBsAg is an excellent system for epitope presentation and mucosal delivery.

2.2.5 TUBERCULOSIS (TB)

TB is primarily caused by *Mycobacterium tuberculosis* and is a leading infectious disease of the developing world. Since the point of microbial entry is the respiratory tract, mucosal delivery of vaccine antigens is clearly the optimal means by which to induce a mucosal immune response. To develop transgenic *Arabidopsis* plants as a delivery system for a TB antigen known as ESAT-6 (early secretory antigenic target), with a molecular weight of 6 kDa. The efficacy of this subunit vaccine has been examined in clinical trials using a mouse model (Rigano and Walmsley, 2005; Rigano et al., 2006).

2.2.6 HUMAN IMMUNODEFICIENCY VIRUS (HIV)

Over 40 million people are currently infected with HIV, with an estimate of 15,000 new HIV infections taking place per day. Of the world's HIV-positive individuals, 65% are located in sub-Saharan Africa. Since HIV is a complex virus and exists in multiple strains, a multicomponent vaccine consisting of several proteins or peptides from different HIV strains may be required to invoke immunity. Both structural and regulatory proteins from HIV have been expressed as possible vaccine proteins in plants (Webster et al., 2005). Proteins derived from the *env* gene are the most frequently used. The *env* gene product is a trimeric complex of surface glycoproteins, including the surface protein gp120 and the transmembrane protein gp41, both of which mediate entry of virus particles into the host cell. Epitopes derived from the *env* gene have been fused with the plant viral capsid proteins, TMV, alfalfa mosaic virus (AlMV), cowpea mosaic virus (CPMV), TBSV, and potato virus (PVX) as part of an epitope presentation system and are discussed in detail in Chapter 4. Full-length and fragments of proteins derived from the *gag* gene, including the capsid protein P24, and the transcriptional activator or Tat protein, have been expressed in tobacco plants.

2.2.7 PLAGUE VACCINE

Plague is caused by *Yersinia pestis*, bacteria that are carried through infected flea bites, direct contact, and by inhalation of infective materials. Bubonic plague is the most common form of plague and is derived from the bite of a flea that fed previously on infected animals. Pneumonic plague causes the greatest amount of mortality and is transmitted by aerosol. It involves the colonization of alveolar spaces within the lung by *Y. pestis* and rapidly leads to pneumonia.

Y. pestis is responsible for the Black Death, which resulted in the mortality of over 30% of the European population. More recently, great interest has been aroused regarding the potential of *Y. pestis* being used as a biological warfare agent. To date, no vaccine that offers strong protection against plague has been identified. It is therefore important to generate a vaccine that is both economical and that can be used for the large-scale immunization of populations.

Antigens derived from *Y. pestis* that have been determined to induce a sufficient immune response against a challenge with live bacteria include the F1 (Fraction 1) antigen, the major capsular protein, and the V antigen, a secreted protein involved in regulating translocation of the bacterial-derived cytotoxic effector proteins into the cytosol of mammalian cells. It would be useful to create a vaccine that would induce mucosal immunity since *Y. pestis* as a biowarfare agent would be delivered to the mucous membrane. Alvarez et al. (2006) designed an F1-V fusion protein that was expressed in transgenic tomato plants. Tomato fruit were then harvested and freeze-dried to standardize the antigen dose. Immunogenicity of this vaccine was determined via oral ingestion using mice that were previously primed subcutaneously. Both mucosal IgA and serum IgG1 responses were elicited.

2.2.8 MEASLES VIRUS

Measles, a highly contagious childhood viral disease, affects the respiratory tract. Over 30 million cases are reported each year, with the majority of fatalities found in developing nations. A plant-derived measles vaccine that could be orally administered would substantially reduce the frequency of measles in areas where refrigeration and syringes are not readily available. Webster et al. (2006) expressed a full-length hemagglutinin protein of measles virus (MV-H) in lettuce, which, upon intranasal and oral vaccination, could generate an MV-specific immune response. An improved immune response was observed when saponins derived from the bark of the quillaja tree were used as a mucosal adjuvant for oral delivery of MV-H (Pickering et al., 2006).

2.2.9 PAPILLOMAVIRUS

Human papillomavirus (HPV) is the causative agent of virtually all cases of cervical cancer and one of the most common of the sexually transmitted diseases. Cervical cancer remains a main cause of cancer-related death in many developing countries. Current vaccines are too expensive and difficult to distribute widely in these countries. In the last decade, noninfectious VLPs of HPV were created and found to induce protective immunity against infection in mice. VLPs can induce virus-neutralizing antibodies, and thus are attractive candidates for prophylactic vaccines against HPV. HPV VLPs are composed of the HPV major capsid protein L1, and have been demonstrated, using animal models, to provide a long-lasting immune response.

Biemelt et al. (2003) developed transgenic potato and tobacco plants expressing the structural protein L1 of HIV-16. Transcript stability was increased by the addition of the TMV translational enhancer omega sequence and by modification to gene codon usage via synthetic gene construction (Chapter 1). Warzecha et al. (2003) showed that VLPs formed in transgenic potato and tobacco could bind to conformationally dependent and HPV genotype-specific neutralizing antibodies. Furthermore, the authors demonstrated that ingestion of transgenic plant material activated a VLP-specific immune memory response that was dependent upon the coadministration of adjuvant.

Further comparative studies of L1 expression were conducted in transgenic plants versus a TMV expression vector. In this case, the cottontail rabbit papillomavirus (CRPV), a close relative of HPV, was used. Rabbits vaccinated with purified protein were protected against wart development upon challenge with live CRPV despite the fact that VLPs were not detected in this animal model system (Kohl et al., 2006).

In addition, mice fed with transgenic potatoes expressing HPV16 VLPs did not evoke a serum antibody response. The response could be enhanced by using a subimmunogenic dose of purified VLPs via injection or oral gavage. High expression of the L1 protein is toxic to plants, and so far, the levels expressed in plants are too low to be used in clinical trial. Nonetheless, the foregoing studies by various research groups provide a valuable step forward in the production of a vaccine against HPV that can be used in developing countries.

2.2.10 SMALLPOX VIRUS

The threat of bioterrorism has brought about a renewed interest in novel approaches to the production of vaccines against smallpox. The current virus-based vaccine displays side effects. The main target used for validation is the BR5 gene, which encodes a virus-specific membrane glycoprotein. High levels of BR5 were found to be expressed in transgenic collard plants in which the BR5 construct was targeted to the apoplast of the cell. Animals that were parenterally immunized were protected against lethal challenge with vaccinia virus (Golovkin et al., 2007).

2.2.11 INFLUENZA VIRUS

Influenza A infections are responsible for 300–500,000 deaths and 3–5 million hospitalizations per year. Every epidemic brings about new strains of influenza A, which arise due to point mutations within the surface glycoproteins hemagglutinin (HA) and neuraminidase (NA). These mutations in turn enable emerging virus strains to evade the host's immune system. The long production time of the current commercially available vaccine, which is produced in chicken eggs, is a major obstacle.

A full-length version of the HA protein (plant-optimized and containing a KDEL endoplasmic reticulum retention signal and a 6His polyhistidine affinity purification tag) was generated in *N. benthamiana* leaves by agroinfection (Shoji et al., 2008). This plant-derived HA was purified by chromatography and determined to be properly folded. The protein was expressed at high levels and it induced serum IgG titers in mice that were comparable to the commercial egg-produced, formalin-inactivated virus. Both hemagglutinin inhibition (HI) and virus neutralizing (VN) antibody titers correlated well with levels observed in mice serum immunized with the commercial virus.

2.3 ANIMAL VACCINES PRODUCED IN PLANTS

Vaccine proteins produced in plants have been developed for animal health and veterinary purposes as well. A number of them are listed here.

2.3.1 FOOT AND MOUTH DISEASE VIRUS

Foot and mouth disease virus (FMDV), an economically important disease affecting meat-producing animals, has in the past been controlled by immunization with inactivated virus. Epitopes critical for the induction of neutralizing antibodies are found on the structural protein VP1, and have been expressed in both prokaryotic and eukaryotic systems. Transgenic alfalfa-expressing VP1 of FMDV were generated, and mice have been either parenterally immunized with leaf extract or received fresh leaves in their diet. A virus-specific immune response to both a synthetic peptide and intact virus particles was observed. Mice challenged with the live virus were protected from the disease. This study indicates once again that forage plants such as alfalfa, commonly used in the diet of domestic animals, can be used as a source of edible vaccines (Dus Santos and Wigdorovitz, 2005).

A more recent study involved transgenic *Stylosanthes guianensis* expressing VP1. Mice were orally immunized by providing the transgenic plant tissue in the form of a desiccated powder to their feed. Mice fed in this fashion developed a virus-specific immune response. These results demonstrate the feasibility of using forage plants to produce vaccine proteins that can then be administered as feedstuff additives (Wang et al., 2008).

2.3.2 BOVINE ROTAVIRUS

Transgenic alfalfa plants have also been used for the development of vaccine proteins against bovine rotavirus (BRV). Diarrhea caused by rotavirus infection is a major cause of death of infants in developing countries. Neonates of cattle and newborn calves are also susceptible. Diarrhea can result from either the presence of rotavirus endotoxin or due to virus-mediated destruction of absorption-efficient enterocytes. By producing antibodies in cows, passive protection could be provided to nursing calves. To this end, transgenic alfalfa plants have been developed that produce chimeric protein consisting of the vaccine peptide eBRV4 epitope fused to the GUS open reading frame. Plants were selected based upon GUS activity. Immunogenicity was assessed by parenteral or oral administration of transgenic leaf extracts in mice. The antibody response elicited was shown to be passively transferred to their offspring, which were demonstrated to be protected against rotavirus challenge. A specific secretory antibody response in immunized mothers was induced using this approach. The infants produced serum antibodies at levels comparable to that of their mothers (Dhama et al., 2008; Dus Santos and Wigdorovitz, 2005).

2.3.3 TRANSMISSIBLE GASTROENTERITIS VIRUS

Transmissible gastroenteritis virus (TGEV) is the causative agent of acute diarrhea of newborn piglets and is associated with a high mortality. Glycoprotein S of TGEV was chosen as a putative antigen as it is highly immunogenic and resistant to degradation in the gut. Although the full-length or the globular domain of TGEV glycoprotein S could not be detected by Western blot analysis in transgenic *Arabidopsis thaliana* plants, antigen purified from these plants and then parenterally inoculated into mice were shown to produce neutralizing antibodies (Lamphear et al., 2004).

2.3.4 INFECTIOUS BURSAL DISEASE VIRUS

Infectious bursal disease virus (IBVD) primarily infects poultry. VP2 of IBVD has been expressed in transgenic *Arabidopsis*. While the serum antibody response in chickens fed with leaf extracts worked less efficiently than commercial vaccine (60% versus 90%) it worked just as efficiently (80%) when used in a booster format along with the commercial vaccine (Wu et al., 2004, 2007). In addition to this, tobacco cell cultures have been used to produce a VP2-based vaccine that worked effectively using the subcutaneous route of administration (Miller et al., 2004).

2.3.5 NEWCASTLE DISEASE VIRUS

Newcastle disease virus (NDV) is a prototype paramyxovirus and can cause high mortality in poultry. The F protein of NDV was expressed in transgenic rice, and immunogenicity was demonstrated in a mouse model (Yang et al., 2007). Recently, plant-made NDV has been licensed for use in animals, making it the first commercial plant-made vaccine available for veterinary use.

2.3.6 RABBIT HEMORRHAGIC DISEASE VIRUS (RHDV)

Rabbit hemorrhagic disease virus (RHDV) causes hemorrhagic syndrome and high mortality in wild rabbit populations. As a consequence, the use of edible plants expressing vaccine proteins against RHDV for the protection of wild populations would be of great benefit (Martin-Alonso et al., 2003). The capsid protein VP60 from RHDV has been produced in transgenic plants. Mice that were fed leaves of plants expressing VP60 were primed to a subimmunogenic single-dose vaccine derived from baculovirus (Gil et al., 2006). This plant-made vaccine has also been tested on the target species, not just the mouse model. In a preliminary small-scale study, a freeze-dried homogenate from potato tubers expressing VP60 was demonstrated to stimulate a protective immune response, and the antigen content remained constant even after several months of storage. Rabbits that were challenged with a virulent strain of the virus were completely protected, while the same virus proved lethal to control, nonvaccinated rabbits (Rice et al., 2005).

2.3.7 CANINE PARVOVIRUS (CPV)

Several studies have been conducted involving the use of CPV proteins or peptides as vaccines in plants. Langeveld et al. (2001) examined the immunogenicity of CPV VP2 expressed in tobacco plants. Mice gavaged with the leaf extract were found to exhibit high serum IgG levels. In a later study, Molina et al. (2004) used tobacco chloroplasts to express the 2L21 peptide (the peptide that confers protection to dogs against challenge by CPV) as part of a fusion protein to CTB or GFP. Mice and rabbits parenterally immunized with leaf extracts from transgenic plants produced high titers of IgG and IgA that were capable of recognizing the VP2 protein.

2.4　THE USE OF ANTIBODIES IN PLANTS AS IMMUNOTHERAPEUTIC AGENTS

A wide range of antibody formats have been expressed in transgenic plants and also shown to work effectively (Table 2.1). Plants are very efficient at producing immunoglobulins because of the great similarity in folding and

TABLE 2.1

Antibody and Antibody Bragments Produced in Plants

Epitope	Antibody Type	Plant System	Application	Reference
Rabies	MAb	Tobacco	Antiviral	Ko et al., 2003
Guy's 13	MAb	Tobacco	Dental caries	Hiatt et al., 1989
Sperm agglutination antigen-1 (SAGA1)	MAb	Tobacco	Spermicidal agent	Xu et al., 2007
HIV	MAb	Tobacco	Antiviral	Webster et al., 2005
HBV	MAb	Tobacco	Antiviral	Yano et al., 2004
HSV	MAb	Soybean	Antiviral	Zeitlin et al., 1998
BR-96	MAb	Soybean	Anticancer therapy	Verch et al., 2004
MuC1 (cancer-associated mucin)	SdAb (single-domain antibody, or diabody)	Tobacco	Anticancer therapy	Ismaili et al., 2007
Tumor-associated Lewis Y oligosaccharide	MAb	Tobacco	Anticolorectal cancer, breast cancer therapy	Brodzik et al., 2006
Alpha-HER2 (epidermal growth factor receptor-2) oncogene	scFv	Tobacco	Anticancer therapy	Galeffi et al., 2005
EGF-R (epidermal growth factor receptor)	MAb	Corn	Anticancer therapy	Ludwig et al., 2004; Rodriguez et al., 2005
CEA (carcinoembryonic antigen)	sdAb	Tobacco	Anticancer therapy	Vaquero et al., 2002

assembly mechanisms between plant and mammal cells (Ma et al., 2004). These include full-size antibodies, Fab fragments, single-chain antibody fragments, and single-chain variable region (scFv) fragments; the choice of which form to use depends on the type of disease and therapy regimen. To produce full-sized antibodies, two promoters are required to express heavy- and light-chain genes and to avoid gene silencing by using the same promoter. Antibodies are the most important biopharmaceutical agents presently in clinical trials. Production of plant-derived IgG antibody was first described in 1989 (Hiatt et al., 1989).

Secretory IgA (sIgA) is the predominant form of immunoglobulin found in mucosal surfaces, where it provides the first line of defense against infectious agents. sIgA is composed of heavy and light chains and exists as a dimer linked by a small polypeptide-joining (J) chain. A fourth polypeptide, which is known as the secretory component (sc), stabilizes the polymeric antibody against proteolysis within the harsh environment of the GI tract. Large quantities of monoclonal sIgAs are required for passive immunization in vivo. Plants offer a system that can produce these antibodies in a quantity that is pharmaceutically useful yet inexpensive.

A plant-produced monoclonal sIgA was generated that mimics the naturally occurring form of antibody in human secretions. This was accomplished by generating four separate transgenic plants that each express a separate component of the sIgA complex. A series of sexual crosses were performed to generate plants in which all four protein chains were expressed simultaneously. Biologically active sIgA was assembled efficiently in these plants to become a major population of total antibody produced by the plant. Further analysis showed that, as in mammalian sIgA assembly, the J chain is required for sc association with the immunoglobulin, suggesting that the nature of association in plants is similar to that found in mammals.

A human clinical trial demonstrating that monoclonal sIgA produced and extracted from transgenic tobacco plants was able to prevent colonization by *Streptococcus mutans* in the mouths of human volunteers for over 4 months with the absence of adverse side effects. Production of a sufficient amount of sIgA for each patient required approximately 1 kg of plant tissue (10–15 mature plants), suggesting that the quantities necessary to immunize large numbers of patients can easily be obtained by producing plants in bulk. Furthermore, studies using hydroponic cultures exhibited a yield of 11.7 µg antibody/g root dry mass/day via secretion of the antibody into the surrounding medium (Drake et al., 2003). Studies showed that passive immunization of plant-synthesized sIgA was as protective and specific as the native monoclonal antibody, and that plant-derived sIgA can survive in the oral cavity for up to 3 days. Similarly, sIgA against glycoprotein B of herpes simplex virus (HSV) produced in transgenic soybean was demonstrated to have comparable levels of anti-HSV activity as monoclonal antibody produced in mammalian cell culture, despite the differences in glycosylation patterns detected between each expression system.

Full antibodies generated in plants have been applied in human health care. For example, the company Agracetus produced human antibodies for injecting cancer patients in corn seed, as well as antibodies against HSV-2 in transgenic soybean. Other applications for antibodies expressed in plants include the production of plant virus coat protein antibodies that interfere with infection. Cytosolic antisera against the CP of artichoke mottled crinkle virus enhanced virus resistance in transgenic tobacco lines. Full-size antibodies or scFv fragments generated against TMV in the plant apoplast, on the other hand, exhibited a lower degree of resistance, demonstrating that the selection of a plant compartment for antibody expression is critical for engineering virus resistance. Since plant virus coat proteins possess a broad structural diversity, the use of these antibodies to combat virus infection becomes restrictive. Instead, the generation of antibodies against the functional domains of viral replicates and movement proteins could provide superior virus-resistant plants. Fungal diseases are currently controlled by the application of chemical fungicides. Recently, antibodies against components such as cutinase have been generated in plants to protect against fungal infection.

The most common antibody fragment expressed in plants is svFv. These antibody fragments are fused to the ER retention signal KDEL to increase protein yield. The major factor that limits accumulation of scFv in plant tissue is protein stability. This can be mitigated to a degree by intracellular targeting.

scFv vaccines consist of hypervariable domains of tumor-specific immunoglobulins, and have been effective in blocking tumor progression in mouse models. In this instance, a signal peptide sequence was fused to the scFv target protein to direct it to the plant secretory pathway. The development of plants that can generate functional monoclonal antibodies will have significant implications for passive immunotherapy by offering an economic means to produce large quantities of antibodies that can be topically applied to prevent infection.

2.4.1 Monoclonal Antibodies Expressed in Plants against Rabies Virus

Rabies affects Southeast Asia and Africa and causes 50–60,000 deaths a year, yet this infectious disease does not receive a great deal of financial support because it is not a major cause of mortality in developed countries.

Treatment involves the application of rabies-specific antibodies that are infiltrated around the bite wound, in addition to immunization. The antibodies provide passive protection and are able to neutralize the virus. For developing countries, these antibodies are both too expensive and difficult to produce in large quantities. In addition to this, there is a dramatic shortage of rabies monoclonal antibodies (MAbs) available on a worldwide basis. The production of inexpensive and safe plant-derived MAbs would be useful in postexposure prophylaxis-global health benefit.

Antirabies human MAbs have been developed in tobacco and demonstrated to exhibit an antirabies virus-neutralizing activity and affinity comparable to its mammalian-derived counterpart (HRIG) (Ko et al., 2003). The immunoglobulin was developed using two different promoters to control expression of LC and HC components of the MAb. In addition to this, a KDEL retention signal and HC were inserted into the antirabies MAb to assist with protein stability and to avoid gene silencing. Further examination of the plant-derived MAbs revealed that selected glycosylation patterns in plants compared with their mammalian counterparts did not affect the MAb's antigen affinity-modified N-glycosylation pattern nor its neutralizing and protective efficacy (Ko and Koprowski, 2005).

2.5   OTHER BIOPHARMACEUTICALS AND THERAPEUTIC AGENTS PRODUCED IN PLANTS

The following represent a few examples of biopharmaceuticals currently expressed in plants. Erythropoietin (Epo), a heavily glycosylated protein, is a principal cytokine involved in the regulation and maintenance of a physiological level of circulating erythrocytes. Recombinant human Epo improves anemia caused by renal failure. The N-linked oligosaccharides are thought to play an important role in Epo activity. Human Epo produced in cultured tobacco cells was found to be unstable in the circulation, possibly because the terminal salicylic acid residue was absent from the N-linked oligosaccharide of the Epo expressed in plants. However, since it functioned quite effectively in vitro, this plant-derived version of Epo can be used as a growth factor in the in vivo propagation of erythrocytes.

In another case, a truncated form of human placental alkaline phosphatase (SEAP), lacking a membrane-anchoring domain, was produced in a rhizosecretion system. SEAP was targeted in a tobacco root culture to the intercellular space through the secretion pathway using the SEAP signal peptide. SEAP comprised as much as 3% of the total root-secreted protein, and compares favorably with some of the highest reported tissue contents of recombinant proteins expressed in plants. The advantage of the rhizosecretion system is that it can be operated continuously without destroying the plant. It represents a simple and cost-effective way to isolate recombinant proteins from a simple hydroponic medium.

Oleosins are a class of seed proteins associated with oil-body membranes in developing and mature embryos. As a simple purification procedure, foreign peptides have been routinely fused with oleosin for the production of foreign proteins in plant seeds. Oleosin fusion facilitates protein purification via cleavage of the fusion protein by an endonuclease, followed by a flotation centrifugation procedure in which the oleosin fusion protein floats to the surface with the oil bodies, thus removing recombinant protein along with the oil-body fraction.

The oleosin fusion procedure was used for the purification of the commercially valuable plant-based blood anticoagulant hirudin in transgenic *Brassica carinata* and *Brassica napus*. Hirudin, a natural protein from the medicinal leech *Hirudo medicinalis*, is superior to other anticoagulants such as heparin. Recombinant hirudin was cleaved from oil-bodies using endoproteinase Factor Xa. Released hirudin was biologically active, as determined by a colorimetric thrombin inhibition assay.

Aprotinin is a serine protease inhibitor and is used as a therapeutic agent. Aprotinin generated in transgenic maize is functionally and biochemically identical to its native counterpart and can be efficiently recovered from seed.

The coding sequences of α- and β-globins of human hemoglobin have been fused to the sequence of the chloroplastic transit peptide of the small subunit of Rubisco. These proteins were then coexpressed in transgenic tobacco plants, resulting in the production of a functional form of tetrameric hemoglobin. The results demonstrate that a complex multimeric protein such as recombinant human hemoglobin can be obtained from tobacco in a functional form.

G-protein-coupled receptors in animals and fungi are part of a huge family of plasma membrane proteins that span the membrane bilayer seven times. Five human muscarinicacetylcholine receptors (MAChRs) were introduced into tobacco, and four of these displayed the correct ligand-binding specificity and kinetics, including correct folding and processing of the receptors. Plants may thus provide a convenient source of such receptors, which are increasingly used for pharmaceutical development. G-proteins are widely expressed in higher plants as well; however, these MAChRs do not appear to regulate endogenous tobacco regulatory pathways.

Other therapeutic proteins produced in plants include leuenkephalin, a neuropeptide produced in *B. napus* and *Arabidopsis* as part of the seed storage protein 2S albumin. The peptide was proteolytically cleaved from the storage protein and recovered by high-performance liquid chromatography (HPLC). In addition, a biologically active form of the human growth hormone somatotropin was produced in tobacco chloroplasts at levels 300-fold higher than a similar gene expressed using a nuclear trans-gene approach, indicating that chloroplasts can also produce pharmaceutical proteins in plants with high efficiency. As another example, colorectal cancer antigen GA733-2 has been generated in transgenic tobacco plants. This plant-derived GA733-2 produced a humoral immune response in mice that was comparable to the antigen produced in insect cell culture (Verch et al., 2004).

Tobacco expressing interleukin-10 is currently being studied as a potential treatment for Crohn's disease. Similarly, trichosanthin (a ribosome inactivator used to inhibit HIV infection) and angiotensin have been expressed in plants. Hirudin (antithrombin) is commercially expressed in plants (Parmenter et al., 1995; Prakash, 1996).

2.5.1 INTERLEUKINS

Interleukin-4 (IL-4), a cytokine, has great potential for the treatment of cancer, and viral and autoimmune diseases, but unfortunately is difficult to produce in high quantities. Transgenic plants offer a potential cost-effective production system for IL-4 expression. In vivo T-cell proliferation assays showed that IL-4 expressed in tobacco plants were able to retain full biological activity. The testing of this plant-derived IL-4 by oral delivery for the treatment of clinical disease is currently under way (Ma et al., 2005).

Interleukin-12, a heterodimeric cytokine produced by dendritic cells, macrophages, and B-cells, and utilized in cancer immunotherapy, has been expressed in tobacco and tomato plants and shown to have similar biological activity to commercially available recombinant IL-12 (Gutierrez-Ortega et al., 2004, 2005). Recently, mouse IL-12 has been produced in transgenic tobacco plant lines and root cultures (Liu et al., 2008). This plant-derived Il-12 underwent signal cleavage and glycosylation processes correctly and demonstrated biological activity comparable to commercially available IL-2 derived from animal cells, confirming again the use of plants as an effective platform for production of proteins involved in therapeutic applications. Similarly, interleukin-18 has been expressed in tobacco and its biological activity confirmed. The level of IL-18 obtained from tobacco plants is sufficient to induce responsiveness in vivo (Zhang et al., 2003).

2.5.2 TAXADIENE

The taxanes are a group of polycyclic diterpenes produced by various species of trees (such as yew) and are most commonly known as the potent anticancer drug paclitaxel (Taxol). The extraction process is costly.

Transgenic tomato fruit expressing taxadiene synthase were able to reroute the production of taxadiene, allowing extraction of large amounts of highly purified taxadiene from freeze-dried tomato (Kovacs et al., 2007).

2.5.3 ALLERGENS

Two recombinant allergens from olive pollen (Ole e 3 and Ole e 8) were produced in *Arabidopsis* and their relative biological activities assessed to be positive via calcium-binding assays. The immunological behavior of these recombinant allergens was equivalent to natural allergens, as demonstrated by their ability to bind to allergen-specific rabbit IgG antiserum and IgE from sensitized patients (Ledesma et al., 2006).

2.5.4 INTERFERON AND OTHER ANTIVIRAL PROTEINS

Interferons are antiviral cytokines, and can also potentiate both innate and adaptive immune response. A number of studies have been conducted using plant-derived antiviral proteins, including interferon. For example, Ohya et al. (2005) have examined the effect of potato-derived IFN-apha to protect mice from infection by *Listeria monocytogenes*. The authors found that the plant-derived IFN was in fact capable of protecting mice at concentrations much lower than that of natural IFN. Similarly, Song et al. (2008) produced chicken alpha interferon (ChIFN-alpha) in transgenic lettuce that exhibited antiviral activity in a VSV-infected chicken embryonic cell line. Furthermore, Arlen et al. (2007) produced IFN alpha2b in tobacco chloroplasts at levels as great as 20% of the total soluble protein.. This chloroplast-derived IFN-alpha2b displayed similar antiviral, antitumor, and immunomodulating activities as its commercial counterpart. Expression of the major histocompatibility complex I (MHC I) and the number of natural killer (NK) cells were increased as well. Tobacco-derived IFN-alpha2b even protected mice from a highly metastatic tumor line.

Microbicides have also been produced in plants and demonstrated to work effectively. For example, cyanovirin-N (CV-N) is able to inactivate a wide range of HIV strains. Tobacco-derived CV-N was shown to be recoverable at 0.85% of total soluble protein in leaf tissue and at 0.64 mg/mL in hydroponic media using rhizosecretion of a hydroponic tobacco culture. CV-N harvested from these plants was demonstrated to be functional in vivo by specific binding to gp120 of HIV, as well as by protecting T-cells from HIV infection (Sexton et al., 2006). In another example, a microbicidal killer toxin of yeast, known as killer peptide (KP), known to work against a variety of human pathogens, has been expressed in plants via a potato virus X expression vector system (Donini et al., 2005). Purified virus particles containing the killer peptide fused to the coat protein were shown to act as effective microbicidal agents against a number of human and phytopathogens.

2.5.5 BIRTH CONTROL

The possum (*Trichosurus vulpecula*) is not a native of New Zealand but was introduced years ago from Australia. Possums have multiplied rapidly and have caused significant economical and environmental damage in New Zealand. One approach to effectively controlling the population is that of immunocontraception. Reproductive antigens for this species have been expressed at high levels in transgenic plants, with the idea that female possums fed plants containing the antigen would exhibit reduced fertility (Pokinghorne et al., 2005). Oral delivery of the immunocontraceptive antigen did indeed reduce fertility; however, further studies are required that will increase efficacy before this method of vaccine delivery can be effectively used in the wild.

Using another approach, a murine monclonal antibody that recognizes sperm agglutination antigen-1 (SAGA1), an antigen present on the surface of sperm cells, has been investigated as a potential spermicidal agent. Transgenic tobacco BY-2 cell lines expressing this antibody have been generated, and the inclusion of a 6-his tag facilitates rapid and inexpensive large-scale purification of this plant-made monoclonal antibody (Xu et al., 2007). This represents yet another use of monoclonal antibodies produced in plants (see Section 2.4).

2.6 ENHANCEMENT OF PLANTS FOR NUTRITIONAL OR MEDICINAL PURPOSES (NUTRICEUTICALS)

The term *nutriceutical* refers to a food extract that has a demonstrable physiological benefit or provides protection against a chronic disease. Examples of foods that have been traditionally considered to be nutriceuticals include flavonoids and antioxidants found in foods as diverse as red grapes, ginseng, and garlic. Advances in plant biotechnology have also brought forth additional or strengthened health benefits to some traditional foods, and several examples are listed here.

2.6.1 GOLDEN RICE

Transgenic plants that possess nutritionally enhanced traits are also currently under development. The most recent example of this is found in the case of "golden rice," an enriched rice that derives its name from its golden hue and was developed by Ingo Potrykus of the Swiss Federal Institute of Technology and Peter Beyer of the Institute of Freiberg.

In Southeast Asia, an estimated 5 million children develop an eye disease known as xerophthalmia due to vitamin A deficiency every year. Of these, 0.25–0.5 million will eventually go blind. Vitamin A deficiency is also correlated with a weakened immune system and consequentially an increased susceptibility to potentially fatal afflictions, including diarrhea, respiratory diseases, and childhood diseases such as measles. According to statistics compiled by UNICEF, improved vitamin A nutrition could be expected to prevent approximately 1–2 million deaths a year among children aged 1–4, and an additional 0.25–0.5 million deaths during later childhood.

Rice represents up to 80% of the daily calorie intake of inhabitants of Southeast Asia. Since the rice endosperm lacks vital micronutrients such as vitamin A, the development of genetically engineered rice strains that produce vitamin A could alleviate the problem. This was accomplished by inserting three genes into rice to allow the plant to produce β-carotene, the precursor required for animals to make vitamin A. This is the first time a trait requiring multiple genes had been transferred into a plant, and included two genes from daffodil and one from the bacterium *Erwina*. The three enzymes added to the rice constitute a complete biosynthetic pathway for β-carotene. The transgenic rice produced golden grains with more than enough β-carotene to meet the daily requirement of vitamin A in meal-sized portions, and even an excess of dietary β-carotene would have no harmful properties to the consumer. Scientists plan to distribute seeds for golden rice free of charge for noncommercial use in developing countries and in rice-breeding programs.

2.6.2 OTHER NUTRICEUTICALS PRODUCED IN PLANTS

Other possible means of increasing nutritional content through the development of transgenic plants have been under examination. A seed-specific amaranth seed albumin (AmA1) from *Amaranthus hypochondriacus* was expressed in transgenic potato tubers. AmA1 is considered to be a good donor protein since it is well balanced in amino acid composition, non-allergenic, and encoded by a single gene. Transgenic plants expressing this protein have an increase in most essential amino acids, thus improving their nutritive value. Expression of this protein in nongrain crops can therefore improve human nutrition.

Carotenoids produced in plants are used as colorants in foods and animal feeds and can also have an antioxidant function. Production of phytoene synthase is the first committed step towards carotenoid biosynthesis in plants. When phytoene synthase is produced in *B. napus,* a 50-fold increase in carotenoid expression results. Therefore *Brassica* and possibly other oil-seed crops may be used as commercial sources of carotenoids.

In addition to enhancement with essential vitamins, amino acids, and proteins, plants can also be metabolically engineered to produce nutritionally superior carbohydrates and lipids. The relative inexpensiveness as well as the capability to grow large-scale quantities make plant production an attractive feature. In the case of carbohydrates such as starch and sucrose, many products or modifications of these products can be produced on a large scale and at much lower costs than are currently available. For example, trehalose, a food additive, was in the past too costly for large-scale production; however, it has now been produced in transgenic tobacco tissue at a much reduced cost.

Transgenic plants have been used to produce improved vegetable oils with novel and modified fatty acids that can be harvested directly from seed. Not only can the fatty acid composition of the oils in transgenic plants be changed but the biosynthesis of the lipid structure itself can be engineered. For example, the stearoyl-acyl-carrier protein (stearoyl-ACP) desaturase catalyzes the initial desaturation reaction in fatty acid biosynthesis and therefore plays an important role in determining the ratio of total saturated to nonsaturated fatty acids in plants. Transgenic plants expressing antisense stearoyl-ACP desaturase RNA in *Brassica rapa* and *B. napus* exhibited a reduction of stearoyl-ACP desaturase protein and a concomitant dramatic increase in higher levels of stearate in mature seed oil. In another study, an acyl–acyl carrier protein thioesterase was expressed in the seeds of transgenic *Brassica* plants. These plants accumulated 68% more stearate than plants that express the wild-type enzyme. In addition, fatty acids that possess conjugated double bonds used in paints, varnishes, and inks were generated in transgenic soybean embryos. Other lipid compounds, such as the biodegradable plastic PHB poly[*R*-(–)-3-hydroxybutyrate], have been synthesized in *Arabidopsis thaliana* plants. In the future, transgenic plants will be generated with improved growth rates, increased photosynthetic efficiency, enhanced specific vitamins, and other biochemical materials such as caffeine or phytic acid removed. These future alterations will serve to improve plants as food sources.

2.6.3 ENHANCING THE NUTRITIONAL CONTENT OF PLANTS BY ALTERATION OF THEIR NATURAL METABOLIC PATHWAYS

There are additional means by which to develop plant varieties with enhanced nutritional functions. For example, recombinant antibodies expressed in plants can be used to inactivate or sequester specific host proteins or compounds, resulting in significant changes in various metabolic pathways. These antibodies can be targeted to either directly interfere by inactivating the enzyme itself or the enzyme's substrate by blocking protein–protein interactions or by changing the secondary structure of an antigen (Nocke et al., 2006). Antibody-based engineering has also been used to alter secondary metabolic pathways to produce compounds of interest. An scFv specific for solasodine glycoside, a secondary metabolite having antiskin carcinoma properties, has been generated in the ER of hairy root cultures. Production of this antibody resulted in increased levels of solasodine glucoside in direct proportion to the level of soluble scFv available in the plant. As another example, plant cytochrome P450 is required for the inexpensive production of antineoplastic drugs such as taxol. Studies involving the alteration of the metabolic pathway of P450 using specific antibodies expressed in plants offers the potential to increase the production of other nutriceuticals such as phytoestrogens and antioxidants (Morant et al., 2003).

2.7 CONCLUSIONS

This chapter illustrated the broad spectrum of uses for plant-derived vaccines and therapeutic proteins. Many of the biopharmaceuticals listed in this chapter were developed in transgenic tobacco or potato plants. While tobacco is not ideal for the expression of vaccine proteins nor is raw potato ideal for oral consumption, they are both relatively easy to work with and have been well characterized, making them useful for proof-of-concept studies. The use of plants for production systems and delivery vehicles holds great promise for future biopharmaceutical development. Proteins can be produced in plants while remaining biologically functional; they can be scaled up for large production and purified inexpensively and with relative ease. The following chapters describe the many attributes of plant-made biopharmaceuticals in more detail.

REFERENCES

Alvarez, M.L., Pinyerd, H.I.., Crisantes, J.D., Rigano, M.M., Pinkhasov, J., Walmsley, A.M., Mason, H.S., and Cardineau, G.A. (2005). Plant-made subunit vaccine against pneumonic and bubonic plague is orally immunogenic in mice. *Vaccine* 24(14): 2477–2490.

Arlen, P.A., Falconer, R., Cherukumilli, S., Cole, A., Cole, A.M., Oishi, K.K., and Daniell, H. (2007). Field production and functional evaluation of chloroplast-derived interferon-alpha2b. *Plant Biotechnol. J.* 5(4): 511–525.

Ashraf, S., Singh, P.K., Yadav, D.K., Shahnawuz, M., Mishra, S., Sawant, S.V., and Tuli, R. (2005). High level expression of surface glycoprotein of rabies virus in tobacco and its immunoprotective activity in mice. *J. Biotechnol.* 119(1): 1–14.

Biemelt, S., Sonnewald, U., Galmbacher, P., Willmitzer, L., and Müller, M. (2003). Production of human papillomavirus type-16 virus-like particles in transgenic plants. *J. Virol.* 77(17): 9211–9220.

Boehm, R. (2007). Bioproduction of therapeutic proteins in the 21st century and the role of plants and plant cells as production platforms. *Ann. N.Y. Acad. Sci.* 1102: 121–134.

Bonis, H.E. (2002). Three criteria for establishing the usefulness of biotechnology for reducing micronutrient malnutrition. *Food Nutr. Bull.* 23(4): 351–353.

Brodzik, R., Glogowska, M., Bandurska, K., Okuliez, M., Deka, D., Ko, K., van der Linden, J., Leusen, J.H., Pogrebnyak, N., Golovkin, M., Steplewski, Z., and Koprowski, H. (2006). Plant-derived anti-Lewis Y MAb exhibits biological activities for immunotherapy against human cancer cells. *Proc. Natl. Acad. Sci. U.S.A.* 103(23): 8804–8809.

Chen, H.F., Chang, M.H., Chiang, B.L., and Jong, S.T. (2006). Oral immunization of mice using transgenic tomato fruit expressing VP1 protein from enterovirus 71. *Vaccine* 24(15): 2944–2951.

Chowdhury, K. and Bagasra, O. (2007). An edible vaccine for malaria using transgenic tomatoes of varying sizes, shapes and colors to carry different antigens. *Med. Hypotheses* 68(1): 22–30.

Cramer, C.L., Boothe, J.G., and Oishi, K.K. (1999). Transgenic plants for therapeutic proteins: linking upstream and downstream strategies. *Curr. Top. Microbiol. Immunol.* 240: 95–118.

De Neva, M., De Buck, S., De Wilde, C., Van Houdi, H., Strobbe, I., Jacobs, A., Van, M., and Epicker, A. (1999). Gene silencing results in instability of antibody production in transgenic plants. *Mol. Gen. Genet.* 260: 582–592.

Dhama, K., Chauban, R.S., Mahendran, M., and Malik, S.V. (2008). Rotavirus diarrhea in bovines and other domestic animals. *Vet. Res. Commun.* 33(1): 1–23.

Dong, J.L., Liang, B.G., Jin, Y.S., Zhyang, W.J., and Wang, T. (2005). Oral immunization with pBsVP6-transgenic alfalfa protects mice against rotavirus infection. *Virology* 339(2): 153–163.

Donini, M., Lico, C., Baschieri, S., Conti, S., Magliani, W., Polonelli, L., and Benvenuto, E. (2005). Production of an engineered killer pepetide in *Nicotiana benthamiana* by using a potato vine virus X expansion system. *Appl. Environ. Microbiol.* 71(10): 6330-6367.

Drake, P.M., Chargelegue, D.M., Vine, N.D., van Dolleweerd, C.J., Obregon, P., and Ma, J.K. (2003). Rhizosecretion of a monoclonal antibody protein complex from transgenic tobacco roots. *Plant Mol. Biol.* 52: 233–241.

Dus Santos, M.J. and Wigdorovitz, A. (2005). Transgenic plants for the production of veterinary vaccines. *Immunol. Cell Biol.* 83: 229–238.

Evangelista, R.L., Kusnadi, A.R., Howard, J.A., and Nikolov, Z.L. (1998). Process and economic evaluation of the extraction and purification of recombinant beta-glucuronidase from transgenic corn. *Biotechnol. Progress* 14: 807–614.

Fischer, R., Schumann, D., Zimmermann, S., Drossard, J., Sack, M., and Schillberg, S. (1999). Expression and characterization of bispecific single-chain Fv fragments produced in transgenic plants. *Eur. J. Biochem.* 262: 810–816.

Fischer, R., Twyman, R.M., and Schillberg, S. (2003). Production of antibodies in plants and their use for global health. *Vaccine* 21(7–8): 820–825.

Galeffi, P., Lombardi, A., Donato, M.D., Latini, A., Sperandei, M., Cantale, C., and Giacomini, P. (2005). Expression of single-chain antibodies in transgenic plants. *Vaccine* 23(15): 1823–1827.

Gaume, A., Komamytsky, S., Borisjuk, N., and Raskin, I. (2003). Rhizosecretion of recombinant proteins from plant hairy roots. *Plant Cell Rep.* 21, 1188–1193.

Gil, F., Titarenko, E., Terrada, E., Arcalis, E., and Escribano, J.M. (2006). Successful oral prime-immunization with VP60 from rabbit haemorrhagic disease virus produced in transgenic plants using different fusion strategies. *Plant Biotechnol. J.* 4(10): 135–143.

Golovkin, M., Spitsin, S., Andrianov, V., Smirnov, Y., Xiao, Y., Pogrebnyak, N., Markley, K., Brodzik, R., Gleba, Y., Isaacs, S., and Koprowski, H. (2007). Smallpox subunit vaccine produced in plants confers protection in mice. *Proc. Natl. Acad. Sci. U.S.A.* 104(16): 6864–6869.

Gu, Q., Han, N., Liu, J., and Zhu, M. (2006). Expression of Heliobacter pylori urease subunit B gene in transgenic rice. *Biotechnol. Lett.* 28(20): 1661–1666.

Gutierrez-Ortega, A., Avila-Moreno, F., Saucedo-Arias, L.J., Sanches-Torres, C., and Gómez-Lim, M.A. (2004). Expression of a single-chain human interleukin-12 gene in transgenic tobacco plants and functional studies. *Biotechnol. Bioeng.* 85(7): 734–740.

Gutierrez-Ortega, A., Sandoval-Montes, C., de Olivera-Flores, T.J., Santos-Argumedo, L., Gomez-Lim, M.A. (2005). Expression of functional interleukin-12 from mouse in transgenic tomato plants. *Transgenic Res.* 14(6): 877–885.

Harzecha, H., Mason, H.S., Lane, C., Tryggvesson, A., Rybicki, E., Williamson, A.-L., Clements, J.D., Rose, R.C. (2003). Oral immunogenicity of human papillomavirus-like particles expressed in potato. *J. Virol.* 77(16): 8702–8711.

Hiatt, A., Caffertey, R., and Bowdish, K. (1989). Production of antibodies in transgenic plants. *Nature* 342: 76–78.

Hood, E.E., Woodard, S.L., and Hon, M.E. (2002). Monoclonal antibody manufacturing in transgenic plants: myths and realities. *Curr. Opin. Biotechnol.* 13(6): 630–635.

Horvitz, M.A., Simon, P.W., and Tanumihardjo, S.A. (2004). Lycopene and beta-carotene are bioavailable from lycopene 'rend' carrots in humans. *Eur. J. Clin. Nutr.* 58(5): 803–811.

Ismaili, A., Jalali-Javaran, M., Rasaee, M.J., Rahbarizadeh, F., Forouzandeh-Maghadam, M., and Memari, H.R. (2007). Production and characterization of anti-(mucin MUC1). single-divalent antibody in tobacco (*Nicotiana tabacum* cultivar Xanthi). *Biotechnol. Appl. Biochem.* 47(1): 11–19.

Jobling, S.A., Jarman, J., The, M.M., Holmberg, N., Blake, C., and Verhoeyen, M.E. (2003). Immunomodulation of enzyme function in plants by single-domain antibody fragments. *Nat. Biotechnol.* 21: 77–80.

Kathuria, S., Sriraman, R., Nath, R., Sack, M., Pal, R., Artsaenko, O., Talwar, G.P., Fiecher, R., and Finnem, R. (2002). Efficacy of plant-produced recombinant antibodies against HCG. *Hum. Reprod.* 17: 2054–2061.

Khoudi, H., Laberge, S., Ferello, J.M., Bazin, R., Darveau, A., Castonguay, Y., Allard, G., Lemieux, R., and Vezina, L.P. (1999). Production of a diagnostic monoclonal antibody in perennial alfalfa plants. *Biotechnol. Bioeng.* 64: 135–143.

Ko, K., Takoah, Y., Rudd, P.M., Harvey, D.J., Dwek, R.A., Spitsin, S., Hanton, C.A., Rupprecht, C., Dietzschold, B., Golovkin, M., and Koprowski, H. (2003). Function and glycosylation of plant-derived antiviral monoclonal antibody. *Proc. Natl. Acad. Sci. U.S.A.* 100: 8013–8018.

Kohl, T., Hitzeroth, I.I., Stewart, D., Varsani, A., Govan, V.A., Christensen, N.D., Williamson, A.-L., and Rybicki, E.P. (2006). Plant-produced cottontail rabbit papillomavirus L1 protein protects against tumor challenge: a proof-of-concept study. *Clin. Vaccine Immunol.* 13(8): 845–853.

Koprowski, H. (2005). Vaccines and sera through plant biotechnology. *Vaccine* 23(15): 1757–1763.

Korban, S.S., Krasnyanski, S.F., and Busetow, D.E. (2002). Foods as production and delivery vehicles for human vaccines. *J. Am. Coll. Nutr.* 21(3 Suppl): 2128–2178.

Kovacs, K., Zhang, L., Linforth, R.S., Whittaker, B., Hayes, C.J., and Fray, R.G. (2007). Redirection of carotenoid metabolism for the efficient production of taxadiene [taxa-4(5), 11(12)-diene] in transgenic tomato fruit. *Transgenic Res.* 15(1): 121–126.

Lai, P., Ramachandran, V.G., Goyal, R., and Sharon, R. (2007). Edible vaccines: current status and future. *Indian J. Med. Microbiol.* 25(2): 93–102.

Lamphear, B.J., Jilka, J.M., Kesl, L., Welter, M., Howard, J.A., and Streatfield, S.J. (2004). A corn-based delivery system for animal vaccines: an oral transmissible gastroenteritis virus vaccine boosts lactogenic immunity in swine. *Vaccine* 22(19): 2420–2424.

Larrick, J.W., Yu, L., Naftzger, C., Jaiswal, S., and Wycoff, K. (2001). Production of secretory IgA antibodies in plants. *Biomol. Eng.* 18: 87–94.

Ledesma, A., Moral, V., Villalba, M., Salinas, J., and Rodriguez, R. (2006). Ca^{2+} binding allergens from olive pollen exhibit biochemical and immunological activity when expressed in stable transgenic Arabidopsis. *FEBS* 273(19): 425–434.

Li, H.Y., Ramalingam, S., and Chye, M.L. (2006). Accumulation of recombinant SARS-CoCV spike protein in plant cytosol and chloroplasts indicate potential for development of plant-derived oral vaccines. *Exp. Biol. Med.* (Maywood) 231(8): 1346–1352.

Li, J.T., Fei, L., Mon, Z.R., Wei, J., Tang, Y., He, H.Y., Wang, L., and Wu, Y.Z. (2006). Immunogenicity of a plant-derived edible rotavirus subunit vaccine transformed over fifty generations. *Virology* 356(1–2): 171–178.

Liu, J., Dolan, M.C., Reidy, M., and Cramer, C.L. (2008). Expression of bioactive single-chain murine IL-12 in transgenic plants. *J. Interferon Cytokine Res.* 28(6): 381–392.

Ludwig, D.L., Witte, L., Hicklin, D.J. Prewett, M., Bassi, R., Burtrum, D., Pereira, D.S., Jimenez, X., Fox, F., Saxena, B., Zhou, Q., Ma, Y., Kang, X., Patel, D., Barry, M., Kussie, P., Zhu, Z., Russell, D.A., Petersen, W.L., Jury, T.P.,

Gaitan-Gaitan, F., Moran, D.L., Delannay, X., Storrs, B.S., Tou, J., Zupec, M.E., Gustafson, K.S., McIntyre, J., Tarnowski, S.J., and Bohlen, P. (2004). Conservation of receptor antagonist anti-tumor activity by epidermal growth factor receptor antibody expressed in transgenic corn seed. *Hum. Antibodies* 13(3): 81–90.

Ma, J. K.-C., Drake, M.W., Chargelegue, D., Obregon, P., and Prada, A. (2004). Antibody processing and engineering in plants, and new strategies for vaccine production. *Vaccine* 23(15): 1814–1818.

Ma, J.K., Drake, P.M., and Christou, P. (2003). The production of recombinant pharmaceutical proteins in plants. *Nat. Rev. Genet.* 4: 794–805.

Ma, J.K., Hiatt, A., Hein, M., Vine, N.D., Wang, F., Stabila, P., van Dolleweerd, C., Mostov, K., and Lehner, T. (1995). Generation and assembly of secretory antibodies in plants. *Science* 258: 716–719.

Ma, J.K., Hikmat, A., Wycoff, K., Vine, N.D., Charelegue, D., Yu, L., Hein, M.B., and Lehner, T. (1998). Characterization of a recombinant plant monoclonal secretory antibody and preventative immunotherapy in humans. *Nat. Med.* 4: 601–606.

Ma, S., Huang, Y., Davis, A., Yin, Z., Mi, Q., Menassa, R., Brandle, J.E., and Jevnikar, A.M. (2005). Production of biologically active human interleukin-4 in transgenic tomato and potato. *Plant Biotechnol. J.* 3(3): 309–318.

Magnuson, N.S., Linzmaier, P.M., Gao, J.W., Reeves, R., An, G., and Lee, J.M. (1996). Enhanced recovery of a secreted mammalian protein from suspension culture of genetically modified tobacco cells. *Protein Exp. Purif.* 7: 220–228.

Martin-Alonso, J.M., Castanon, S., Alonso, P., Parra, F., and Ordas, R. (2003). Oral immunization using tuber extracts from transgenic potato plants expressing rabbit hemorrhagic disease virus capsid protein. *Transgenic Res.* 12: 127–130.

Mayfield, S.P., Franklin, S.E., and Lerner, R.A. (2003). Expression and assembly of a fully active antibody in algae. *Proc. Natl. Acad. Sci. U.S.A.* 100: 438–442.

Miller, T., Fanton, M., and Webb, S. (2004). Stable immunoprophylactic and therapeutic compositions derived from transgenic plant cells and methods for production. World Patent Application WO 2004/098530.

Mishra, H.N. and Das, C. (2003). A review on biological control and metabolism of aflatoxin. *Crit. Rev. Food Sci. Nutr.* 43(3): 245–264.

Molidrem, K.L., Li, J., Simon, P.W., and Tanumihardjo, S.A. (2004). Lutein and beta-carotene from lutein-containing yellow carrots are bioavailable in humans. *Am. J. Clin. Nutrition* 80(1): 131–136.

Molina, A., Hervás-Stubbs, S., Daniell, H., Mingo-Castel, A.M., and Veramendi J. (2004). High-yield expression of a viral peptide animal vaccine in transgenic tobacco chloroplasts. *Plant Biotechnol J.* 2(2): 141–153.

Molina, A., Veramendi, J., and Hervás-Stubbs, S. (2005). Induction of neutralizing antibodies by a tobacco chloroplast-derived vaccine based on a B cell epitope from canine parvovirus. *Virology* 342(2): 266–275.

Morant, M., Bak, S., Moller, H.L., and Woerck-Reichhart, D. (2003). Plant cyto-chromes P450: tools for pharmacology, plant protection, and phytoremedia-tion. *Curr. Opin. Biotechnol.* 14(22): 151–162.

Moravec, T., Schmidt, M.A., Herman, E.M., and Woodford-Thomas, T. (2007). Production of *Escherichia coli* heat labile toxin (LT) B subunit in soybean seed and analysis of its immunogenicity as an oral vaccine. *Vaccine* 25(9): 1647–1657.

Nolke, G., Fischer, R., and Schillberg, S. (2006). Antibody-based metabolic engi-neering in plants. *J. Biotechnol.* 124(1): 271–283.

Ohya, K., Matsumura, T., Itchoda, N., Ohashi, K., Onuma, M., and Sugimoto, C. (2005). Ability of orally administered IFN-alpha-containing transgenic potato extracts to inhibit *Listeria monocyogenes* infection. *J. Interferon. Cytokine Res.* 25(8): 429–466.

Parmenter, D.L., Boothe, J.G.H., van Rooijen, G.J. Yeung, E.C., and Maloney, M.M. (1995). Production of biologically active hirudin in plant seeds using oleosin partitioning. *Plant Mol. Biol.* 29(6): 1167-1180.

Peeters, K., De Wilde, C., De Jaeger, G., Angenon, G., and Depicker, A. (2001). Production of antibodies and antibody fragments in plants. *Vaccine* 19: 2756–2761.

Peeters, K., De Wilde, C., and Depicker, A. (2001). Highly efficient targeting and accumulation of a F(ab) fragment within the secretory pathway and apoplast of *Arabidopsis thaliana. Eur. J. Biochem.* 268: 4251–4260.

Pickering, R.J., Smith, S.D., Strugnell, R.A., Wesselingh, S.L., and Webster, D.E. (2006). Crude saponins improve the immune response to an oral plant-made measles vaccine. *Vaccine* 24: 144–150.

Polkinghorne, I., Hamerili, D., Cowan, P., and Duckworth, J. (2005). Plant-based immunocontraceptive control of wildlife—"potentials, limitations and pos-sums." *Vaccine* 23(15): 1847–1850.

Rice, J., Ainley, W.M., and Shewen, P. (2005). Plant-made vaccines: biotech-nology and immunology in animal health. *Animal Health Res. Rev.* 6(2): 199–209.

Rigano, M.M. and Walmsely, A.M. (2005). Expression systems and developments in plant-made vaccines. *Immunol. Cell Biol.* 83: 271–277.

Rigano, M.M., Dreitz, S., Kipnis, A,-P,, Izzo, A.A., and Walmsley, A.M. (2006). Oral immunogenicity of a plant-made tuberculosis vaccine. *Vaccine* 24(6): 691–695.

Rodriguez, M., Ramirez, N.I., Ayala, M., Freyre, F., Perez, L., Triguero, A., Mateo, C., Selman-Housein, G., Gavilondo, J.V., and Pujol, M. (2005). Transient expression in tobacco leaves of an aglycosylated recombinant antibody against the epidermal growth factor receptor. *Biotechnol. Bioeng.* 89(2): 188–194.

Santi, L., Huang, Z., and Mason, H. (2006). Virus-like particles production in green plants. *Methods* 40(1): 66–76.

Schillberg, S., Fischer, R., and Emans, N. (2003). Molecular farming of recombi-nant antibodies in plants. *Cell Mol. Life. Sci.* 60: 433–445.

Schillberg, S., Zimmermann, S., Voss, A., and Fischer, R. (1999). Apoplastic and cytosolic expression of full-size antibodies and antibody fragments in Nicotiana tabacum. *Transgenic Res.* 8: 255–263.

Schillberg, S., Fischer, R., and Emans, N. (2003). Molecular farming of recombinant antibodies in plants. *Cell Mol. Life Sci.* 60(3): 433–445.

Schillberg, S., Fischer, R., and Emans, N. (2003). "Molecular farming" of antibodies in plants. *Naturwissenschaften* 90(4): 145–155.

Schouten, A., Roosien, J., Bakker, J., and Schots, A. (2002). Formation of disulfide bridges by a single-chain Fv antibody in the reducing ectopic environment of the plant cytosol. *J. Biol. Chem.* 277: 19339–19345.

Sexton, A., Drake, P.M., Mahmod, N., Harman, S.J., Shattock, R.J., and Ma, J.K-C. (2005). Transgenic plant production of Cyanovirin-N, an HIV microbicide. *FASEB J.* December, express article 10.1096/fj.05-4742fje

Sharp, J.M. and Doran, P.M. (2001). Strategies for enhancing monoclonal antibody accumulation in plant cell and organ cultures. *Biotechnol. Progress* 17: 979–992.

Shoji, Y., Chichester, J.A., Bi, H., Musiychuk, K., de la Rosa, P., Goldschmidt, L., Horsey, A., Ugulava, N., Palmer, G.A., Mett, V., and Yusibov, V. (2008). Plant-expressed HA as a seasonal influenza vaccine candidate. *Vaccine* 26 (23): 2930–2934.

Song, L. , Zhao, D.G., Wu, Y.J., and Li, Y. (2008). Transient expression of chicken alpha interferon gene in lettuce. *J. Zhejiang Univ. Sci.* B 9(5): 351–355.

Streatfield, S.J. (2006). Mucosal immunization using recombinant plant-based oral vaccines. *Methods* 38(2): 150–157.

Stoger, E., Vaquero, C., Torres, E., Sack, M., Nicholson, L., Drossard, J., Williams, S., Keen, D., Perris, Y., Christou, P., and Fischer, R. (2000). Cereal crops as viable production and storage systems for pharmaceutical scFv antibodies. *Plant Mol. Biol.* 42: 583–590.

Stoger, E., Sack, M., Fischer, R., and Christou, P. (2002). Plantibodies: applications, advantages and bottlenecks. *Curr. Opin. Biotechnol.* 13: 161–166.

Tekoah, Y., Ko, H., Koprowski, H., Harvey, D.J., Wormald, M.R., Dwek, R.A., and Rudd, P.M. (2004). Controlled glycosylation of therapeutic antibodies in plants. *Arch. Biochem. Biophys.* 426: 266–278.

Vaquero, C., Sack, M., Schutster, F., Finnern, R., Drossard, J., Schumann, D., Reimann, A., and Fischer, R. (2002). A carcinoembryonic antigen-specific diabody produced in tobacco. *FASEB J.* 16(3): 408–410.

Verch, T., Hooger, D.C., Kiyathin, A., Stephewski, Z., and Koprowski, H. (2004). Immunization with a plant-produced colorectal cancer antigen. *Cancer Immunol. Immunother.* 53(2): 92–99.

Wang, D.M., Zhu, J.B. Peng, M., and Zhou, P. (2008). Induction of a protective antibody response to FMDV in mice following oral immunization with transgenic *Stylosanthes* spp. as a feedstuff additive. *Transgenic Res.* (Epub ahead of print).

Webster, D.E., Smith, S.D., Pickering, R.J., Strugnell, R.A., Dry, I.B., and Wesselingh, S.L. (2006). Measles virus hemagglutinin protein expressed in transgenic lettuce induces neutralizing antibodies in mice following mucosal vaccination. *Vaccine* 24(17): 3538–3544.

Webster, D.E., Thomas, M.C., Pickering, R., Whyte, A., Dry, I.B., Gorry, P.R., and Wesselingh, S.L. (2005). Is there a role for plant-made vaccines in the prevention of HIV/AIDS? *Immunol. Cell Bio.* 83(3): 239–247.

Wu, H., Singh, N.K., Locy, R.D., Scissum-Gunn, K., and Giambrone, J.J. (2004). Immunization of chickens with VP2 protein of infectious bursal disease virus expressed in *Arabidopsis thaliana*. *Avian Dis.* 48: 663–668.

Wu, J., Yu, L., Long, L., Hu, J., Zhou, J., and Zhou, X. (2007). Oral immunization with transgenic rice seeds expressing VP2 protein of infectious bursal disease virus induces protective immune responses in chickens. *Plant Biotechnol. J.* 5(5): 570-578.

Wycoff, K.L. (2005). Secretory IgA antibodies from plants. *Curr. Pharm. Des.* 11(19): 2429–2437.

Xu, B., Copolla, M., Herr, J.C., and Timko, M.P. (2007). Expression of a recombinant human sperm-agglutinating mini-antibody in tobacco (*Nicotiana tabacum* L.). *Soc. Reprod. Fertil. Suppl.* 63: b465–477.

Yang, Z.Q., Liu, Q.Q., Pan, Z.M., Yu, H.X., and Jiao, X.A. (2007). Expression of the fusion glycoprotein of Newcastle disease virus in transgenic rice and its immunogenicity in mice. *Vaccine* 25(4): 591–598.

Yano, A., Maeda, F., and Takekoshi, M. (2004). Transgenic tobacco cells producing the human monoclonal antibody to hepatitis B virus surface antigen. *J. Med. Virol.* 73: 208–215.

Zhang, B., Yang, Y.H., Lin, Y.M., Rao, Q., Zheng, G.G., and Wu, K.F. (2003). Expression and production of bioactive human interleukin-18 in tobacco plants. *Biotechnol. Lett.* 19: 1625–1635.

Zhang, H., Zhang, X., Liu, M., Zhang, J., Li, Y., and Zheng, C.C. (2006). Expression and characterization of *Helicobacter pylori* heat-shock protein A (HspA) protein in transgenic tobacco (*Nicotiana tabacum*) plants. *Biotechnol. Appl. Biochem.* 43 (Pt.1): 33–38.

Zhang, X., Buehner, N.A., Hutson, A.M., Estes, M.K., and Mason, H.S. (2006). Tomato is a highly effective vehicle for expression and oral immunization with Norwalk virus capsid protein. *Plant Biotechnol. J.* 4(4): 419–432.

Zeitlin, L., Olmsted, S.S., Moench, T.R., Co, M.S., Martinell, B.J., Paradkar, V.M., Russell, D.R., Queen, C., Cone, R., and Whaley, K.J. (1998). A humanized monoclonal antibody produced in transgenic plants for immunoprotection of the vagina against genital herpes. *Nat. Biotechnol.* 16: 1361–1364.

3 Chloroplast Engineering and Production of Biopharmaceuticals

3.1 INTRODUCTION

One of the most promising plant expression systems for protein production stems from the engineering of chloroplasts via plastid transformation. Chloroplast engineering involves the introduction of thousands of copies of a foreign gene of interest per cell, and, as a consequence, the production of the foreign protein may reach levels as great as 46% of total soluble protein (TSP) (Daniell, 1999; Daniell et al., 2002; Daniell and Dhingra, 2002). This feature gives chloroplast engineering enormous promise for the production of vaccine proteins and biopharmaceuticals. In addition, plastid transformation is a more environmentally friendly approach to plant engineering because it both minimizes outcrossing of transgenes to weedy relatives and eliminates the possibility of transferring potentially toxic transgenic pollen to nontarget insects, as will be discussed in more depth in the following text.

Chloroplast engineering originated in the 1980s, initially by plastid isolation and then by the introduction of transformed plastids into protoplasts and regeneration of transgenic plants (Bock, 2001). Later, with the development of the gene gun, it became possible to transform chloroplasts without plastid isolation. The first chloroplast transformations were performed on *Chlamydomonas reinhardtii* cells. Plant cells themselves have evolved from the original capture of a cyanobacterium by a pre-eukaryotic, mitochondria-possessing cell. The majority (>3000) of genes of the endosymbiont were either lost or transported to the nucleus due to the gradual integration of the acquired endosymbionts into the host cell's metabolism (Bock and Khan, 2004).

The plastid genome of modern higher plants remains well conserved, with little interspecific variation in genomic organization and coding capacity. Identical copies of the plastid genome are present in a diverse array of plastid transformation types: proplastids (precursor plastids found in most plant cells and present predominantly in meristematic tissues), green chloroplasts (present in photosynthetically active tissues such as leaves, immature fruit, and stems), carotenoid-accumulating red or yellow chromoplasts (present in mature fruit and flowers), amyloplasts (present in starch storage organs), leucoplasts (present in flowers, roots, and nongreen stems), elaioplasts (present in fat or oil storage organs of the mature plant), and etioplasts (partially developed chloroplasts found in dark-grown seedlings) (Bock, 2001; Table 3.1).

Each chloroplast genome contains 120–130 genes in a densely packed, circular double-stranded DNA molecule of 120–160 kb in length. In spite of their small size, plastid DNA makes up as much as 10%–20% of total cellular DNA content since a single plant cell contains thousands of copies of its plastid genome. Chloroplast DNA is organized into nucleoids in a fashion typical of a prokaryotic system. Several such nucleoids are present in each chloroplast, and each nucleoid harbors several copies of the plastid genome. The ploidy levels can therefore be extremely high; for example, up to 10,000 identical copies of plastid DNA can be found in a single pea leaf cell, and up to 50,000 copies of plastid DNA can be found in a single wheat cell.

Plastids have an active homologous recombination system and, as a result, the genome exists in a continuous state of inter- and intramolecular exchange (Heifetz, 2000). Furthermore, numerous prokaryotic features have been retained in plastids; for example, plastid genes are organized into operons and, in general, are coexpressed to form polycistronic mRNAs. Other prokaryotic-like mechanisms of gene expression also exist within plastids. In addition to this, plastids possess highly sophisticated

TABLE 3.1
Plastids and Their Characteristics

Plastid	Characteristics
Proplastids	Small, undifferentiated plastids found in most plant cells, predominate in meristematic tissues
Amyoplasts	Storage organs for starches
Chloroplasts	Center for photosynthesis, present in leaves, immature fruit, and stems
Chromoplasts	Contain colored lipids, accumulate yellow or red carotenoids; present in mature flowers and fruit
Etioplasts	Specific stage in the transformation of proplastids to chloroplasts, occur in plant tissues grown in the absence of light
Elaioplasts	Storage organs for fats and oils
Leucoplasts	Synthesize lipids and other materials; colorless, found in flowers, roots, and stems

regulatory mechanisms that govern the cooperation of plant cell organelles within their nucleocytoplasmic compartment (Heifetz, 2000).

3.2 DIFFERENCES BETWEEN NUCLEAR AND PLASTID TRANSFORMATIONS

Plastids possess a number of unique features that make them attractive systems for the production of biopharmaceuticals. For example, as mentioned earlier, transformed chloroplasts may contain thousands of copies of the transgene of interest per cell, whereas only one or a few copies of a transgene are generated per cell during a nuclear transformation. As a result of this, the level of accumulation of the corresponding protein as a percentage of total soluble protein is much greater in transformed plastids than in transformed nuclei. Insertion of foreign DNA into the plastid genome is site specific, and multiple genes can be inserted simultaneously. Insertion into the nuclear genome, on the other hand, is undirected and, in general, only single genes are inserted one at a time. Expression of chloroplast-derived transgenes does not undergo any position effects or exhibit gene silencing as nuclear-derived transgenes do. Chloroplast transformants are unable to outcross because of the absence of chloroplast DNA in pollen. This provides such transformants with a natural gene-containment strategy. These features and others that distinguish plastid from nuclear transformation are illustrated in Table 3.2.

3.3 PLASTID TRANSFORMATION

3.3.1 BASIC CONCEPT

Plastid transformation in plants is based on the introduction of foreign DNA by biolistic delivery (bombardment with a particle gun) or PEG (polyethylene glycol) treatment, followed by homologous recombination between the plastid-targeting sequences of the transformation vector and the targeted region of the plastid genome (Bock and Khan, 2004). A plastid transformation vector consists of two targeting sequences of approximately 400 bp with homology to the chloroplast genome that together flank the foreign gene of interest. A RecA-type system of homologous transformation results in the insertion of the gene of interest at a precisely predetermined location in the chloroplast genome at a high level of efficiency, resulting in uniform expression among transgenic lines (Maliga, 2003).

A few years ago, genetic transformation of chloroplasts seemed impossible to achieve because of two significant challenges. First, the double membrane of chloroplasts offered a significant barrier; no viruses or bacteria were known to infect chloroplasts that could be used as vectors for

TABLE 3.2
Differences Between Nuclear and Plastid Transformations

Feature	Chloroplast	Nucleus
Transgene copy number	Thousands of copies	Few copies
Level of gene expression	Up to 47% TSP[a]	Much lower
Insertion of DNA into genome	Site-specific insertion	Multiple insertions, undirected
Transcription of genes	Multiple genes arranged as operons in a single transformation event	Each gene expressed individually
Position effect, gene silencing	No effects	Variable, decreased transgene expression
Gene containment	Natural gene containment due to maternal inheritance	Potential outcrossing among weeds
Protein folding	Disulfide bonds formed, no glycosylation	Disulfide bond formation, glycosylation (plant-specific pathways)
Toxicity of protein	Chloroplast compartmentalization minimizes adverse effects of protein	May have serious effects in cytoplasm
Homozygosity	Transgenic lines are homoplasmic	Can be either hetero- or homozygous
Major limitation	Only a few crop species successfully transformed; many are infertile	Variability and low level of protein expression

[a] TSP = total soluble protein.

gene transfer. This was circumvented by the invention of the "Gene Gun," which is used to introduce foreign DNA into plastids via biolistic delivery. Other mechanisms of transformation, such as PEG treatment or micro-injection, have also been employed and are described in more detail in the following text. Since biolistic delivery is less time consuming and less demanding with respect to tissue culture technology, it is the more widely used transformation procedure at present.

The other barrier to transformation technology was the challenge of uniformly altering the enormous copy number of the plastid genome, a fundamental requirement for genetic stability. If a homoplasmic genome is not achieved, rapid somatic segregation will then take place, resulting in genetic instability. For plastids, the primary transformation event involves the change of a single or small number of plastid genome copies within a single chloroplast out of the 10,000 or so plastid DNA (ptDNA) copies present in the average leaf mesophyll cell. Primary transplastomic cell lines therefore contain a mixed population of wild-type and transformed plastid genomes; these heteroplastomic situations are genetically unstable and resolve spontaneously into either one of two types of homoplasmy due to random segregation upon both organelle division and cell division. By allowing for a sufficient number of cell divisions under a high selective pressure, homoplastomic shoots can be obtained after two to four such cycles of regeneration. In this way, chloroplasts that exclusively harbor wild-type genomes are sensitive to the selecting antibiotic and will not multiply efficiently. Plastids that possess wild-type genomes eventually disappear (Figure 3.1). As a result, the problem of heteroplasmy can be circumvented by repeated rounds of selection and plant regeneration in tissue

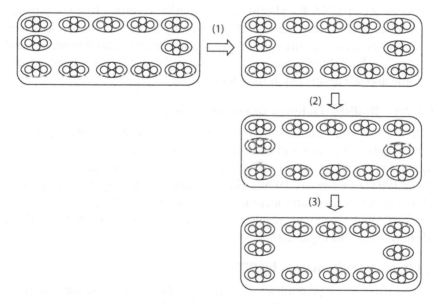

FIGURE 3.1 *A color version of this figure follows page 110.* Establishment of a transformed, homoplasmic cell line. The primary chloroplast transformation event involves the change of a single copy of the plastid genome out of thousands of copies in a single leaf cell. Stages (1)–(3) each represent cell divisions under selective antibody pressure. After each cell and organelle division, the selective antibiotic favors the multiplication of those chloroplasts that contain the transformed copies of the genome. At stage (2), some chloroplasts may contain a mixed population of transformed and wild-type genomes (heteroplasmy). At stage (3), those chloroplasts that harbor wild-type genomes are eventually eliminated, and homoplasmy is achieved over several rounds of plant regeneration on selective medium.

culture; this permits the elimination of many residual wild-type genomes that are present in the primary transformants (Bock, 2001).

The selection of transformed chloroplasts usually involves the use of an antibiotic resistance marker. Spectinomycin is used most routinely because of the high specificity it displays as a prokaryotic translational inhibitor as well as the relatively low side effects it exerts on plants. The bacterial aminoglycoside 3′-adenyltransferase gene (*aad*A) confers resistance to both streptomycin and spectinomycin. The aadA protein catalyzes the covalent transfer of an adenosine monophosphate (AMP) residue from adenosine triphosphate (ATP) to spectinomycin, thereby converting the antibiotic into an inactive form that no longer inhibits protein synthesis for prokaryotic 70S ribosomes that are present in the chloroplast.

3.3.2   PLASTID TRANSFORMATION SYSTEMS

The delivery of DNA for plastid transformation requires first the passage of genetic material through the cell wall and plasma membrane, followed by a further crossing through the organelle's double membrane. Several approaches have been employed to deliver foreign DNA to the plastid genome and are described in the following text.

3.3.2.1   Biolistic Delivery of Foreign DNA

In this approach, micron- or submicron-sized tungsten or gold particles coated with foreign DNA to be used for transformation are accelerated to a high velocity and impacted into cultured cells using a Gene Gun (Sanford et al., 1993; Finer et al., 1999). The advantage of this system is that it is relatively inexpensive and technically very easy. The disadvantage is that the force of delivery can result in mechanical damage or sheared DNA (Heifetz, 2000; Maliga, 2003).

3.3.2.2   Transformation by PEG

Transformation of protoplasts by PEG has been demonstrated to work well for chloroplasts as well as plant nuclei. Since chloroplasts tend to be appressed to the cell membranes in protoplasts, they can be penetrated easily by this method. However, this approach is technically more demanding than biolistic delivery and involves a considerably longer period of time for the selection process (Heifetz, 2000).

3.3.2.3   Microinjection

Another approach that has been explored for plastid transformation is microinjection. Microinjection into discrete plastids of intact plant cells entails the use of a syringe consisting of a submicron diameter glass capillary driven by the thermal expansion of galinstan, a liquid metal alloy

(Heifetz, 2000). Although protoplasts that have been transformed in this way have been shown to exhibit transient gene expression, stable plastid transformation has not yet been successfully achieved (Knoblauch et al., 1999).

3.3.2.4   DNA Transfer Mediated by T-DNA of *Agrobacterium tumefaciens* and Other Approaches

The first attempt at plastid transformation employed a sophisticated approach using isolated chloroplasts and an *Agrobacterium* binary transformation vector. This approach, however, has not yet been demonstrated to work successfully in plants (Block et al., 1985; Ward et al., 2002). A number of other approaches for plastid transformation are currently under development. One of these involves the introduction of foreign DNA into isolated chloroplasts, followed by the incorporation of the transformed organelles into protoplasts. This approach has not yet been shown to work effectively (Daniell and McFadden, 1987; Daniell and Dhingra, 2002). Furthermore, as an alternative to stable integration, "shuttle vectors," or vectors with a plastid *ori* (origin of replication) sequence, have been episomally maintained in plastids when plants that have been transformed in this manner were grown on selective media (Staub and Maliga, 1994, 1995). Although this approach has been shown to be successful, shuttle vectors rapidly become lost in the absence of selection.

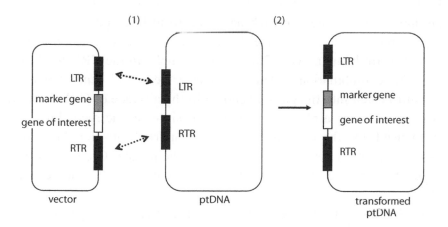

FIGURE 3.2   Design of vector and targeting of gene of interest to a specific region of the plastid genome by site-directed homologous recombination. Flanking regions LTR and RTR represent left and right targeting regions, respectively. (1) Plasmid integration occurs by two homologous recombination events in the flanking regions (dashed double-ended arrows). (2) As a result, the plastid genome acquires the selectable marker gene (e.g., *aadA*) and the gene of interest.

3.3.3   Vectors and Selectable Markers Used

Plastid vectors have left and right plastid-targeting regions, known as LTR and RTR, which are required for homologous recombination (Figure 3.2). Targeting regions have homology to the chosen targeting site and span 1–2 kb in length. The promoter most frequently used for plastid transformation is the strong plastid rRNA operon (rrn) promoter. Transformation vectors currently contain the *aad*A gene that encodes aminoglycoside 3′-adenyltransferase, an enzyme that inactivates spectinomycin and streptomycin by adenylation. Transplastomic clones can then be identified as shoots on spectinomycin media (Maliga, 2003). The neomycin phosphotransferase gene *npt*II, which confers resistance to kanamycin and other aminoglycoside antibiotics, has also been used instead of *aad*A, with less success (Svab and Maliga, 1993; Heifetz, 2000).

A system for the elimination of marker genes was also developed so that marker genes (of which only a few are available) could be reused for other transformation events. The removal of marker genes for selection also addresses regulatory concerns regarding the release of antibiotic resistance genes in field crops. Since spectinomycin and streptomycin are commonly used to combat bacterial infection in humans and animals, there has also been some concern that their overuse might lead to the development of a resistant strain of bacteria (Daniell and Dhingra, 2002). One approach has been to utilize endogenous chloroplast recombinases to delete the marker gene via engineered direct repeats (Iamtham and Day, 2000). For example, in the CRE-*loxP* site-specific recombination system, the marker gene and gene of interest are introduced into the plastid genome in the absence of CRE activity. A gene encoding a plastid-targeted CRE site-specific recombinase is introduced into the nucleus that can be imported into the plastid and excise sequences between the *loxP* sites. The nuclear Cre gene is subsequently removed by segregation in the seed progeny (Corneille et al., 2001).

In the future, it may be feasible to restrict transgene expression to a particular tissue or developmental stage by placing the transgene under phage T7 RNA polymerase promoter control. Plastid transgene expression can be switched on by a nuclear-encoded and plastid-targeted T7 RNA polymerase. Expression of T7 RNA polymerase can in turn be controlled by tissue-specific, developmental stage-specific, or chemically inducible promoters (Heifetz, 2000).

3.3.4   Advantages of and Stumbling Blocks
to Plastid Transformation

3.3.4.1   Advantages

There are many advantages to utilizing plastids for transformation; some of them are listed in the following sections.

3.3.4.1.1   Polyploidy Number

Since thousands of copies of the plastid genome can be found per plant cell, extremely high levels of foreign protein are able to accumulate in plants that harbor transgenic chloroplasts. Expression levels greater than 40% of total soluble cellular protein, or 10–100 times higher than protein expression from nuclear transformants, have been detected from plastid transformants.

3.3.4.1.2   Containment of Transgenes and Environmental Safety

Plastid genes are inherited uniparentally in a strictly maternal fashion. This can be a consequence of plastid exclusion by unequal cell division during pollen grain mitosis. Alternatively, degradation of plastids and plastid DNA may take place during male gametophyte development. Since plastid DNA is lost during the process of pollen maturation, it is not carried on to the next generation (Bock and Khan, 2004). This eliminates the probability of uncontrolled contamination of nearby fields with pollen from transgenic crops and the possibility of outcrossing of pollen by gene flow from transgenic crops to related wild species (Maliga, 2003). Thus, transgenic plastid technology results in a much greater level of environmental safety than is found for transgenic nuclear genome manipulations.

3.3.4.1.3   Toxicity of Foreign Protein

Since plastids have a limited set of protein degradation pathways, foreign proteins that exhibit harmful effects to the plant in the cytoplasm may be more stable when they accumulate within the chloroplast (Heifetz, 2000). For example, the vaccine protein cholera toxin B subunit was shown to be toxic even when it accumulated to very low levels within the plant cytoplasm, but was nontoxic when it accumulated to large quantities within the chloroplast. Plastids also possess the ability to form disulfide bonds, a requirement for many correctly folded mammalian proteins (Daniell et al., 2005a). These properties have made them attractive for the production of biopharmaceuticals in plants.

3.3.4.1.4   Lack of Position or Gene Silencing Effects

Chloroplast transgenes are integrated in a site-specific manner via homologous recombination. As a result, all transplastomic transformants are genetically and phenotypically identical. This eliminates the problem of positional effects due to random integration of transgenes into unpredictable locations through nonhomologous transformation. Positional effects often pose a significant problem for nuclear transformed plants, resulting in widely varying expression levels. Instead, transgene expression is more stable and uniform among transgenic chloroplast lines. In addition, nuclear transformants frequently suffer from epigenetic gene silencing, which causes variability and reduction in gene expression levels in a given plant over time. Chloroplast transformants are also not affected by epigenetic gene silencing mechanisms as are plants with transformed nuclei (Bogorad, 2000).

3.3.4.1.5   Transgene Stacking

Chloroplast transformations enable multiple genes to be engineered into plants in a single transformation event for simultaneous expression of multiple transgenes. This contrasts with nuclear transformants, in which only the first cistron is normally translated from a polycistronic mRNA. In chloroplasts and prokaryotic systems, multiple cistrons are concurrently expressed from operons. This opens the possibility for the expression of several components of immunoglobulin heavy and light chains or multicomponent vaccine proteins to be generated in a single transformation event. Since the plastid transformation system is versatile, bacterial genes and human cDNAs can be expressed without the need for codon modification (Maliga, 2003; Daniell et al., 2005b).

3.3.4.2   Limitations to Chloroplast Engineering

A number of significant stumbling blocks have been identified that limit the practicality of plastid transformations; the most noteworthy are mentioned here.

3.3.4.2.1   Limitations in Plant Range

Until very recently, high-efficiency chloroplast transformations have been limited to *Chlamydomonas* and the higher plant *Tobaccum*. *Arabidopsis* chloroplasts have been successfully transformed, but none of the regenerated transformed plants were fertile. Chloroplasts from potato leaf cells and rice suspension cells have also been transformed but with poor efficiency as well, and no fertile plants were recovered (Khan et al., 1999; Hibbard et al., 1998). Therefore, a major hurdle is the development of fertile, transgenic, economically important plants.

Since plastid transformation has up until very recently been efficient only in tobacco, there have been a number of obstacles to extending this technology to crop plants that regenerate through somatic embryogenesis. These include (1) tissue culture and regeneration protocols, (2) a lack of versatile selectable markers, and (3) the inability to produce transgenes in nongreen plastids, such as proplastids, amyloplasts, and chromoplasts. Some advances have recently been made in this area (see the following text).

3.3.4.2.2   Plastid Stability

One of the technical challenges of plastid transformation is the problem of ensuring that all the transgene-containing chromosomes are not lost, and to obtain and maintain genetically stable homoplasmic chloroplast transformants. Because each cell may have over 50 plastids, and each chloroplast has approximately 60 chromosomes, there are over 3000 plastid chromosomes per cell. It is difficult to ensure that none of these are wild-type chromosomes that may preferentially replicate in the absence of selectable markers. One of the reasons why efficient transformation has been limited to tobacco is that tobacco cells maintain photosynthetically active chloroplasts during in vitro culture, and the plastid genome of tobacco is naturally sensitive to antibiotics that affect 70S ribosomes such as spectinomycin. The ribosomal RNA of many monocotyledonous plants, however, is resistant to this antibiotic. Thus, plants of this type would have difficulty retaining transformed plastids. It is not realistic to maintain plants perpetually on selective media. Better selectable markers will therefore be a requirement for transformation of many agronomically important crops (Heifetz, 2000).

3.3.4.2.3   Target Organelle for Foreign Protein Accumulation

Transformation of plastids may be disadvantageous if the protein product is required to leave the plastid and go to another part of the plant cell. Unfortunately, little is known about the export of macromolecules from this organelle at present.

3.3.5   TRANSFORMATION IN OTHER CROPS

Plastid transformation is highly dependent on the tissue culture process because it enables copies of the wild-type plastid genome to be selectively eliminated before plant regeneration (Maliga, 2003). However, many of the crop species regenerated in this way turn out to be sterile, a consequence of plant regeneration from tissue culture. As mentioned earlier, the transformation of *Arabidopsis*, tomato, potato, rice, and rape seed oil has been achieved at very low efficiencies, and the resulting transformants were found to be infertile. Thus, it appears that plastid transformation in different taxonomic groups requires different approaches. The greatest challenge has been the manipulation of a somatic embryogenesis system for each crop type. One of the reasons for this low efficiency has been that the vectors used for transformation of some crops, such as potato or tomato, used flanking sequences from tobacco or *Arabidopsis*. As a result, species-specific vectors must be used to expand the number of crops available for transformation. Recently, the first successful demonstrations of plastid genetic engineering by somatic embryogenesis using carrot, cotton, and soybean crop plants have taken place (Daniell et al., 2005a).

3.4   HISTORY OF PLASTID TRANSFORMATION AND BIOTECHNOLOGICAL APPLICATIONS

Successful chloroplast transformation was first achieved for *Chlamydomonas reinhardtii*, a unicellular green algae with a single large chloroplast occupying ~60% of the cell volume. Using photosynthetically incompetent mutants that lacked chloroplast ATP synthase activity (harboring defective alleles of the chloroplast *atp*B gene), researchers were able to complement mutants with a wild-type *atp*B gene under restored photoautotrophic growth selection and obtain stable transformants via homologous recombination. The first transformation of chloroplasts in a higher plant was achieved in tobacco and involved selection for spectinomycin-resistant plant cell lines in tissue culture, followed by regeneration into mature transgenic plants (Maliga, 2004). A series of reciprocal crosses that took place between transplastomic and wild-type plants demonstrated that the introduced antibiotic-resistance trait was maternally inherited in a manner that would be expected for an extranuclear trait.

The first biotechnological application of chloroplast engineering was the expression of the *Bacillus thuringiensis* (Bt) toxin gene from the tobacco plastid genome; this transformation event resulted in plants that displayed high levels of resistance to herbivorous insects (McBride et al., 1995). Other agronomically valuable traits, such as resistance to fungal and bacterial diseases, drought, and herbicides, have also been produced in transformed plastids. The first human therapeutic protein expressed in chloroplasts was somatotropin and is discussed in more detail in the following text (Staub et al., 2000). The first marker gene elimination, involving loop out by homologous recombination and cotransformation, was performed by Fischer et al. (1996).

More recent biotechnological applications of chloroplast engineering involve the manipulation of plastid DNA sequences in vitro and the reintroduction of altered sequences into the chloroplast genome. The result is the development of a powerful tool for in vivo studies of plastid gene expression, including the *cis*-acting elements involved in transcription, RNA metabolism, and translation (Staub and Maliga, 1994). The availability of transgenic technologies for chloroplast research has facilitated the functional characterization of plastid genome-encoded genes and open reading frames. Using an approach known as reverse genetics for example, inactivation of a known gene with unknown function through mutation may result in the recovery of an analyzable phenotype and the revelation of the gene's function (Quesada-Vargas et al., 2005).

3.5 VACCINES AND THERAPEUTIC PROTEINS PRODUCED IN CHLOROPLASTS

From the aforementioned studies, there is little surprise that chloroplast engineering is also considered to be a promising new means by which to produce vaccines, antibodies, and other therapeutic proteins at high levels in plants. A number of examples are described in the following text and in Table 3.3 (Daniell, 2009).

3.5.1 HUMAN SOMATOTROPIN (HST)

The first human therapeutic protein expressed in plant chloroplasts was produced by Staub et al. (2000), who investigated whether human somatotropin (hST) could be expressed in a biologically active and correctly disulfide-bonded form in tobacco chloroplasts. Somatotropin, which is not glycosylated in mammals, is used to treat hypopituitary dwarfism in children, Turner syndrome, chronic renal failure, and HIV wasting syndrome, among other conditions. The authors designed a chimeric hST gene, containing the yeast ubiquitin open reading frame fused in frame to the hST open reading frame. Cleavage of this ubiquitin–hST fusion protein resulted in the creation of an N-terminal phenylalanine, as is naturally found in the pituitary gland upon removal of the signal peptide from native somatotropin. This chimeric gene was then introduced into tobacco leaves using biolistic delivery (Staub et al., 2000). Transformed shoots were selected by growth on spectinomycin plates. Homoplasmic plants were obtained after two rounds of plant regeneration, and Southern blot analysis was used to verify that these plants were indeed plastid transformants. Results from Western blot analysis indicated that hST expression reached levels as great as 7% of total soluble protein (TSP) in mature leaves selected from tissue culture-derived plants (about sevenfold higher than hST expressed in tobacco via nuclear transformation). These transplastomic tobacco lines expressing high levels

TABLE 3.3
Vaccine and Therapeutic Proteins Expressed in Plastids

Protein	Construct	Promoter	Expression Level	Reference
Human somatotropin	hST	Prrn	7% TSP	Kumar and Daniell, 2004
Cholera toxin	CTB	Prrn	4%	Daniell et al., 2001
Antimicrobial peptide	MSI-99	Prrn	21%	DeGray et al., 2001
Interferon α2b	INFα2B	Prrn	19%	Kumar and Daniell, 2004
Anthrax PA	Pag	Prrn	18.1%	Koya et al., 2005; Azthar Aziz et al., 2005
Canine parvovirus	CTB-2L21	Prrn	31.1%	Molina et al., 2004, 2005
	GFP-2L21	Prrn	22.6%	
Tetanus toxin	Tet C	Prrn	25%	Tregoning et al., 2003
Human serum albumin	hsa	PpsbA	11%	Fernandez-San Millan et al., 2003
Plague vaccine	CaF1-LcrV	Prrn	4.6%	Daniell, 2005b
Gamma interferon	IFN-γ	PpsbA	6%	Leelavathi and Reddy, 2003
Insulin-like growth factor	IGF-1	Prrn	33%	Daniell, 2005b

of hST were not only phenotypically normal upon development but were also fertile. The results of this study demonstrated for the first time the feasibility of using plastids for the expression of human proteins.

3.5.2 CHOLERA TOXIN (CT) AND *E. COLI* HEAT-LABILE TOXIN (LT)

Vibrio cholera, the bacterium that causes cholera, colonizes the small intestine and produces cholera enterotoxin, which, as a consequence, brings about various symptoms including watery diarrhea. Cholera toxin is made up of a single A subunit of 27 kDa (CTA) and a nontoxic pentamer of B subunits of 11.6 kDa (CTB); together, these form a hexameric multimer (Daniell et al., 2001). The A subunit has ADP ribosyl-transferase activity and facilitates entry of the toxin into epithelial cells. When administered orally, cholera toxin B (CTB) is a powerful mucosal immunogen. This is because of its ability to bind to epithelial cell surfaces via GM1 ganglioside receptors present on the intestinal cell surface and to elicit a mucosal immune response to pathogens. This immune response can be enhanced when CTB is chemically coupled to other antigens (Daniell et al., 2001).

CTB has been used as a candidate for oral vaccine development. A study by Daniell et al. (2001) was performed to determine the ability of CTB to be expressed in tobacco chloroplasts. The vector used in these studies contained chloroplast border sequences within its transformation vector and the corresponding homologous sequences of the chloroplast genome. Homoplasmy was achieved within the first round of selection due to the presence of a complete chloroplast origin of replication within the flanking sequence, consequently doubling the copy number of foreign genes per cell. This chloroplast-derived CTB was determined to have a strong affinity for GM1 gangliosides, indicating that it retained the antigenic sites necessary for CTB pentamer binding to the pentasaccharide GM1. The retention of the GM1-binding ability also suggests the correct folding of CTB molecules resulting in a functional pentameric structure. Biochemical analysis of chloroplast-derived CTB indicated that the protein was unusually stable, possibly due to its conditions of storage within chloroplasts, a property that is advantageous to the production of edible vaccines. The authors also determined that CTB expression levels from leaves at mature stages of development possess the highest levels of CTB protein expression (Daniell et al., 2001).

Enterotoxigenic *E. coli* (ETEC) causes severe diarrhea as a result of the production of one or more enterotoxins such as LT, a molecule that is immunologically related to CT. Both LT and CT bind through specific interactions via the B subunit pentamer, with the GM1 ganglioside present on the intestinal epithelial cell surface, and can function as adjuvants as well as immunogens. Heat-labile toxin B (LTB) was expressed at levels as high as 4.1% of TSP in tobacco chloroplasts, a level 410-fold higher than that of native LTB expressed in the nuclear genome. Chloroplast-derived CTB was also able to bind to intestinal membrane GM1-ganglioside receptors, indicating that correct folding and disulfide bond formation of CTB pentamers can take place. Plants expressing LTB could not be distinguished from wild-type plants with respect to growth rates, seed setting, and flowering (Kang et al., 2003).

3.5.3 TETANUS TOXIN

Another vaccine protein that has recently been produced in tobacco chloroplasts is the Fragment C domain of tetanus toxin (TetC). TetC is a nontoxic 47 kDa polypeptide that can induce an immune response upon parenteral immunization. In a pioneering study, Tregoning et al. (2003) designed both a bacterial-derived version of TetC with a high AC content and a synthetic version of TetC with a high GC content. Both were equally stable when expressed in chloroplasts and accumulated to significantly high levels (25% of TSP for bacterial and 10% for synthetic TetC). Mice immunized intranasally with tobacco leaf extracts and plastid-produced TetC were able to induce protective levels of TetC antibodies, making the development of this system for the production of a plant-based mucosal vaccine feasible.

3.5.4 CANINE PARVOVIRUS

Canine parvovirus (CPV) infects young animals and produces hemorrhagic gastroenteritis and myocarditis. To design a peptide-based vaccine against the VP2 protein of CPV that is expressed in chloroplasts, Molina et al. (2004) used a linear antigenic peptide (named 2L21) corresponding to amino acids 2–23 of the VP2 CP. The 2L21 synthetic peptide was chemically coupled to the KLH carrier protein and demonstrated to protect dogs and minks against infection by parvovirus. The 2L21 synthetic peptide has been fused to GUS and expressed in the nuclei of transgenic plants. In this study, Molina et al. (2005) expressed the 2L21 peptide alone or as a fusion protein to either CTB or the green fluorescent protein (GFP) in chloroplasts. Large differences were observed in the accumulation of recombinant proteins from each of the three constructs used in these experiments. For example, plants transformed with the 2L21 construct alone did not produce any detectable peptide based on enzyme-linked immunosorbent assay (ELISA). On the other hand, plants transformed with the CTB-2L21 or GFP-2L21 expressed their fusion protein products at high levels. These high levels of expression of CTB-2L21 or GFP-2L21 did not adversely affect the phenotype of transgenic plants and were indistinguishable from the control, nontransformed plants. The highest protein levels were found in young leaves. Northern blot analysis revealed that plants expressing all three constructs produced transcripts of the expected size and level of abundance so that the low levels found for peptide 2L21 were not the result of deficient transcription or poor mRNA stability. The authors hypothesized that the 22-amino-acid peptide produced may be recognized as a foreign molecule in the stroma of the chloroplast and consequentially would become rapidly degraded, possibly due to a primary structure that is more sensitive to chloroplast peptidases.

The results of this study indicated that chloroplast-synthesized CTB-2L21 protein displays a strong affinity for GM1 gangliosides in a similar manner to that found for purified bacterial CTB. This indicates that the fusion protein retained the correct pentameric structure characteristic of CTB and also conserved the specific antigen-binding sites for the GM1-ganglioside receptors on the intestinal epithelial cells. In addition, Balb/c mice immunized intraperitoneally with leaf extracts from transgenic plants expressing CTB-2L21 elicited an anti-2L21 antibody response. None of the mice immunized with 2L21 recognized the peptide, and only one of the GFP-2L21 group showed a marginal response against 2L21. To examine the humoral response, sera were titrated against VP2 protein. Only mice that had been immunized with tobacco leaf extracts expressing CTB-2L21 recognized VP2 in ELISA. The results of this study suggest that this chimeric fusion protein might be able to induce a protective immune response against CPV.

3.5.5 ANTHRAX PROTECTIVE ANTIGEN (PA)

Anthrax, a disease caused by infection by *Bacillus anthracis* via spores, can be transmitted to humans or animals; ruminants such as sheep, goats, cattle, and deer are most susceptible. The handling of infected animals or animal products may also lead to human infection. Recently, anthrax has been considered to be a potential candidate for bioterrorism activity. The spores are extremely hardy and may come into contact with humans through a cut or abrasion, through consumption of infected meat, or by inhalation. The Center for Disease Control (CDC) lists anthrax as a category A disease, and the only vaccine that currently exists has a number of drawbacks and health risks.

Watson et al. (2004) expressed the immunogenic protective antigen (PA), the primary immunogen of anthrax, in transgenic tobacco chloroplasts so that large quantities of PA vaccine antigen can rapidly be obtained to stockpile in case of an emergency terrorist attack. The authors developed plants that accumulated levels as great as 2.7% of total soluble protein in mature leaves. The functionality of chloroplast-derived PA was proven by its ability to internalize the lethal factor and bring about lysis in a mouse macrophage cytotoxic assay. More detail on chloroplast-derived anthrax and its ability to evoke a mucosal immune response is found in Chapter 7.

3.5.6 HEPATITIS E ORF2 PROTEIN

Hepatitis E virus (HEV), a 27–30 nm RNA virus, is the causative agent of this waterborne disease that is prevalent in many tropical and subtropical areas. As HEV accounts for >50% of acute viral hepatitis in young adults and has a 20% fatality rate in pregnant women, it consequently is a serious threat to public health. Zhou et al. (2006) produced ORF2-derived pE2 peptide of HEV from the chloroplast genome of tobacco plants and found that pE2 not only could be produced in quantities as high as 13.27 μg/g in fresh tobacco leaves but also displayed significant levels of immunogenicity in mice.

3.6 CONCLUSIONS

Although a number of hurdles are yet to be overcome such as limitations in plant range and plastid stability, chloroplasts remain attractive candidates for use in the production of plant-made biopharmaceuticals. The increasing ease with which the chloroplast genome can be manipulated as a result of new technologies such as reverse genetics offers even greater opportunities for the utilization of plastids in biotechnological applications such as the production of edible vaccines. Preliminary studies have indicated that significantly higher levels of vaccine protein are produced in plastids than in nuclei, thus demonstrating the enormous potential of this organelle in further applied research. There is no doubt that emerging plastid technologies will assist in the production of plant-derived biopharmaceuticals in years to come.

REFERENCES

Azthar Aziz, M., Sikriwal, D., Jarugula, S., Kumar, P.A., and Bhatnagar, R. (2005). Transformation of an edible crop with the pagA gene of *Bacillus anthracis*. *FASEB J.* 10: 1096–2031.

Block, M. D., Schell, J., and Montagu, M.V. (1985). Chloroplast transformation by *Agrobacterium tumefaciens*. *EMBO J.* 4: 1367–1372.

Bock, R. (2001). Transgenic plastids in basic research and plant biotechnology. *J. Mol. Biol.* 312: 425–438.

Bock, R. and Khan, M.S. (2004). Taming plastids for a green future. *Trends Biotechnol.* 22(6): 311–318.

Bogorad, L. (2000). Engineering chloroplasts: an alternative site for foreign genes, proteins, reactions and products. *Trends Biotechnol.* 18: 257–263.

Chebolu, S. and Daniell, H. (2009). Chloroplast-derived vaccine antigens and biopharmaceuticals: expression, folding, assembly, and functionality. *Curr. Top. Microbiol. Immunol.* 332: 33-54.

Corneille, S. et al. (2001). Efficient elimination of selectable marker genes from the plastid genome by the CRE-lox site-specific recombination system. *Plant J.* 27: 171–178.

Daniell, H. (1999). GM crops: public perception and scientific solutions. *Trends Plant Sci.* 4(12): 467–469.

Daniell, H., Chebolu, S., Kumar, S., Singleton, M., and Falconer, R. (2005a). Chloroplast-derived vaccine antigens and other therapeutic proteins. *Vaccine* 23: 1779–1783.

Daniell, H. and Dhingra, A. (2002). Multigene engineering: dawn of an exciting new era in biotechnology. *Curr. Opin. Biotechnol.* 13: 1326–1341.

Daniell, H., Khan, M.S., and Allison, L. (2002). Milestones in chloroplast genetic engineering: an environmentally friendly era in biotechnology. *Trends Plant Sci.* 76: 84–91.

Daniell, H., Kumar, S., and Dufourmantel, N. (2005a). Breakthrough in chloroplast genetic engineering of agronomically important crops. *Trends Biotechnol.* 23(5): 238–245.

Daniell, H., Lee, S.-B., Panchal, T., and Wiebe, P.O. (2001). Expression of the native cholera toxin B subunit gene and assembly as functional oligomers in transgenic tobacco chloroplasts. *J. Mol. Biol.* 311: 1001–1009.

Daniell, H. and McFadden, B.A. (1987). Uptake and expression of bacterial and cyanobacterial genes by isolated cucumber etioplasts. *Proc. Natl. Acad. Sci U.S.A.* 84: 6349–6353.

DeGray, G., Rajusokaran, K., Smith, F., Sanford, J., and Daniell, H. (2001). Expression of an antimicrobial peptide via the chloroplast genome to control phytopathogenic bacteria and fungi. *Plant Physiol.* 127: 852-862.

Fernandez-San Millan, A., Mingo-Castel, A., Miller, M., and Daniell, H. (2003). A chloroplast transgenic approach to hyperexpress and purify human serum albumin, a protein highly susceptible to proteolytic degradation. *Plant Biotechnol. J.* 1: 71-77.

Finer, J.J., Finer, K.R., and Ponappa, T. (1999). Particle bombardment mediated transformation. *Curr. Top. Microbiol. Immunol.* 240: 59-80.

Fischer, N. et al. (1996). Selectable marker recycling in the chloroplast. *Mol. Gen. Genet.* 251: 373–380.

Heifetz, P.B. (2000). Genetic engineering of the chloroplast. *Biochimie* 82: 655–666.

Hibbard, J.M. et al. (1998). Transient expression of green fluorescent protein in various plastid types following microprojectile bombardment. *Plant J.* 16: 627–632.

Iamtham, S. and Day, A. (2000). Removal of antibiotic resistance genes from transgenic tobacco plastids. *Nat. Biotechnol.* 18: 1172–1176.

Kang, T.-J., Loe, N.-H., Jang, M.-O., Jang, Y.-S., Kim, Y.-S., Sco, J.-E., and Yang, M.-S. (2003). Expression of the B subunit of *E. coli* heat-labile enterotoxin in the chloroplasts of plants and its characterization. *Transgenic Res.* 12: 683–691.

Khan, M.S. et al. (1999). Fluorescent antibiotic resistance marker for tracking plastid transformation in higher plants. *Nat. Biotechnol.* 17: 910–915.

Knoblauch, M. et al. (1999). A galinstan expansion femotosyringe for micro-injection of eukaryotic organelles and prokaryotes. *Nat. Biotechnol.* 17: 906–909.

Koya, V., Moayeri, M., Leppla, S.H., and Daniell, H. (2005). Plant-based vaccine: mice immunized with chloroplast-derived anthrax protective antigen survive anthrax lethal toxin challenge. *Infect. Immun.* 73(12): 8266–8274.

Kumar, S. and Daniell, H. (2004). Engineering the chloroplast genome for hyper-expression of human therapeutic antigens. In *Recombinant Gene Expression*. Methods Molecular Biology, volume 267. New York: Springer.

Leelavathi, S. and Reddy, V.S. (2003). Chloroplast expression of His-tagged GUS fusion: a general strategy to overproduce and purify foreign proteins using transplastomic plants as bioreactors. *Mol. Breed.* 11: 49-58.

Maliga, P. (2003). Progress towards commercialization of plastid transformation technology. *Trends Biotechnol.* 21(1): 20–28.

Maliga, P. (2004). Plastid transformation in higher plants. *Annu. Rev. Plant Biol.* 55: 289-313.

McBride, K.E. et al. (1995). Amplification of a chimeric Bacillus gene in chloroplasts leads to an extraordinary level of an insecticidal protein in tobacco. *Biotechnology* (NY) 13: 362–365.

Molina, A., Hervds-Stubbs, S., Daniell, H., Mingo-Castel, A.M., and Veramendi, J. (2004). High-yield expression of a viral peptide animal vaccine in transgenic tobacco chloroplasts. *Plant Biotechnol. J.* 2: 141–153.

Molina, A. Veramendi, J., and Hervas-Stubbs, S. (2005). Induction of neutralizing antibodies by a tobacco chloroplast-derived vaccine based on a B cell epitope from canine parvovirus. *Virology* 342: 2666–2675.

Quesada-Vargas, T., Ruiz, O.N., and Daniell, H. (2005). Characterization of heterologous multigene operons in transgenic chloroplasts, transcription processing and translation. *Plant Physiol.* 138: 1746–1762.

Ruf, S., Hermann, M., Berger, I.J., Carrer, H., and Bock, R. (2001). Stable genetic transformation of tomato plastids and expression of a foreign protein in fruit. *Nat. Biotechnol.* 18: 870–875.

Staub, J.M., Garcia, B., Graves, J., Hajdukiewicz, P.T.J., Hunter, P., Nehra, N., Paradkar, V., Schlittler, M., Carroll, J.A., Spatola, L., Ward, D., Ve, G., and Russell, D.A. (2000). High yield production of a human therapeutic protein in tobacco chloroplasts. *Nat. Biotechnol.* 18: 333–338.

Staub, J.M. and Maliga, P. (1994). Extrachromosomal elements in tobacco plastids. *Proc. Natl. Acad. Sci. U.S.A.* 91: 7468–7472.

Staub, J.M. and Maliga, P. (1995). Marker rescue from the *Nicotiana tabacum* plastid genome using a plastid *Escherichia coli* shuttle vector. *Mol. Gen. Genet.* 249: 37–42.

Svab, Z. and Maliga, P. (1993). High-frequency plastid transformation in tobacco by selection for a chimeric aadA gene. *Proc. Natl. Acad. Sci. U.S.A.* 90: 913–917.

Tregoning, J.S., Nixon, P., Kuroda, H., Svab, Z., Clare, S., Bowe, F., Fairweather, N., Ytterberg, J., van Wijl, K., Dougan, G., and Matiga, P. (2003). Expression of tetanus toxin Fragment C in tobacco chloroplasts. *Nucleic Acids Res.* 31(4): 1174–1179.

Ward, D.V. et al. (2002). Agrobacterium VirE2 gets the VIP1 treatment in plant nuclear import. *Trends Plant Sci.* 7: 1–3.

Watson, J., Koya, V., Leppia, S.H., and Daniell, H. (2004). Expression of *Bacillus anthracis* protective antigen in transgenic chloroplasts of tobacco, a non-food/feed crop. *Vaccine* 22: 4374–4384.

Zhou, Y.-X., Lee, M., Ng, J. M.-H., Chye, M.-L., Yip, W.-K., and Lam, E. (2006). A truncated hepatitis E virus ORF2 protein expressed in tobacco plastids is immunogenic in mice. *World J. Gastroenterol.* 14: 12(2): 306–312.

4 Plant Viral Expression Vectors and Production of Biopharmaceuticals in Plants

4.1 INTRODUCTION

The stable transformation of plants has been used routinely as a method to express vaccine and therapeutic proteins. A number of disadvantages such as the length of time taken for transgenic plants to be generated and containment (i.e., preventing the escape of transgenes into the environment) have brought about a search for alternative methods by which to express proteins in plants. One such alternative is the utilization of plant virus expression systems.

Early in the history of plant virology, the first conception of engineering plant viruses as expression vectors for foreign protein production was initiated when (+)-sense RNA plant viruses were found to reprogram infected cells to produce their own proteins encoded by viral genetic material. The prospect of converting these viral cDNA clones into foreign gene expression vectors began once several full-length infectious cDNA clone versions of RNA viruses had been engineered in the mid-1980s (Scholthof et al., 1996; LaComme et al., 2001; Pogue et al., 2002). Initially, plant expression vectors were comprised of recombinant viruses that had been constructed with the foreign gene of interest replacing the CP genes; however, the absence of the CP limited their biological capabilities and rendered many of them unable to move both systemically or via cell to cell through plants. As this field of research progressed, vectors were constructed that were capable of expressing a foreign gene in addition to all of their required viral genes. For example, in the case of many RNA viruses that produce subgenomic RNAs to express their downstream cistrons, an additional subgenomic promoter can be inserted to enable the expression of the foreign gene of interest. In other instances, as in the case of potyviruses, the foreign gene of interest may be cloned in frame with an existing viral open reading frame (ORF), most often the viral CP, or as part of a polyprotein ORF by creating additional proteinase cleavage sites to flank the foreign protein. These fusion proteins can then undergo proteolytic cleavage, and the result is the expression of the foreign protein as a free protein. Furthermore, specific targeting signals can be placed within the protein itself to direct the protein to certain areas of the cell by designing plant viral expression systems that express free foreign proteins. In some instances, this can produce higher accumulation by directing the protein to alternative subcellular compartments (Scholthof et al., 1996).

4.2   PLANT RNA VIRUS EXPRESSION SYSTEMS

Two types of expression systems based on plant RNA viruses have been developed for production of immunogenic peptides and proteins in plants: epitope presentation systems (short antigenic peptides fused to the CP that are displayed on the surface of assembled viral particles) and polypeptide expression systems (these systems express the whole unfused recombinant protein that accumulates within the plant).

4.2.1   Epitope Presentation Systems

Insertion sites are chosen in this expression strategy so that the peptide is displayed on the surface of the virus particle without interfering with the ability of the modified CP to assemble into mature virions. As a result, this system has been developed mainly for viruses in which the topology of the capsid protein has been resolved to some degree of detail (Figure 4.1) (Fitchen et al., 1995; Johnson et al., 1997; Fernandez-Fernandez et al., 1998).

4.2.1.1   Cowpea Mosaic Virus

Cowpea mosaic virus (CPMV), a bipartite RNA virus, was the first plant virus to be developed as an epitope presentation system. CPMV has a well-characterized structure comprised of 60 copies each of both large (L) and small (S) coat protein subunits that are arranged to form an icosahedral virus capsid surrounding a bipartite RNA genome. The first identification of exposed loops on the virus surface was properties that made CPMV suitable as a virus expression vector (Lomonossoff and Johnson, 1996; Lomonossoff and Hamilton, 1999). The βB-βC loop of the S protein is the most exposed loop, and this loop has been extensively used as the site of insertion of foreign peptides. Both the βE-αB and βC'-βC" loops have also been used for this purpose (Charterji et al., 2002; Porta et al., 2003). In order to maintain stability of the virus particle, an inserted peptide can be no more than 40 amino acids in length and must have a pI of

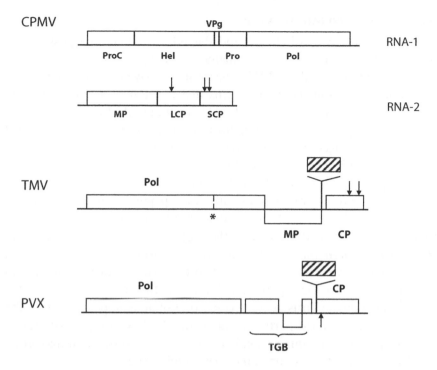

FIGURE 4.1   Genome organization of several plant viruses used to express heterologous peptides/proteins in plants. Positions where epitopes have been inserted within the capsid protein of each virus are indicated by arrows. Positions where a foreign gene of interest has been inserted into the virus genome is indicated by hatched boxes. ProC: proteinase cofactor; Hel: helicase; VPg: virus genome-linked protein; Pro: protease; Pol: polymerase; MP: movement protein; LCP: large capsid protein; SCP: small capsid protein; TGB: triple gene block; CPMV: cowpea mosaic virus; TMV: tobacco mosaic virus; PVX: potato virus X. (Adapted from Canizares et al. (2005). *Immunol. Cell Biol.* 83: 263–270.)

less than 9.0 in the CPMV antigen presentation system (Bendahamane et al., 1999). Under these restrictions, modified particles have virus yields similar to those of wild-type CPMV (Lomonossoff and Johnson, 1996). Many epitopes that have been displayed on this virus were able to successfully elicit strong immune responses. A number of these CPMV-based chimeras have been purified and injected into animals, which were then demonstrated to raise antibodies against the inserted peptide. Antigen presentation systems for vaccine use will be discussed in more detail in a later section. Immunological assays utilizing a number of CPMV chimeras have been used in various immunological assays. In several cases, including vaccine peptides corresponding to canine parvovirus, mink enteritis virus and *P. auruginosa* protective immunity has been observed.

4.2.1.2   Tobacco Mosaic Virus

Tobacco mosaic virus (TMV), a rod-shaped virus that also possesses a 6.5 kbp RNA genome, is the type member of the genus *Tobamovirus* and encodes four polypeptides. TMV is the most extensively studied plant RNA virus and, as a result, is another virus that is routinely used for antigen presentation. The virus particle of TMV contains 2130 copies of capsid protein, making it an attractive platform for peptide display. TMV capsid proteins are arranged in a helical symmetry. There is little space available on the virus surface for insertion of a relatively large foreign protein due to the tight packing together of the CP subunits.

Since there are size restrictions, only small peptide epitopes can be displayed on TMV particles. Epitopes that have been expressed using TMV vectors include those derived from the foot and mouth disease virus and *Pseudomonas aeroginisa*. Initial experiments using TMV as an antigen display system were unsuccessful due to the inability of CP subunits to assemble into stable virus particles. To mitigate this hurdle, a second ORF was included on the virus genome consisting of an epitope containing the coat protein gene. Virus particles that contain both wild-type coat protein and coat protein containing an epitope were designed by the introduction of a leaky termination codon at the coat protein gene.

Another strategy to avoid the problem of small size restriction was employed by Yusibov et al. (2002). In this approach, the CP gene of alfalfa mosaic virus (AlMV) and containing the desired epitope was expressed under the control of the TMV CP subgenomic promoter in such a way that both TMV and AlMV virus particles could be produced simultaneously. Epitopes from rabies virus and human immunodeficiency virus (HIV) have been expressed on AlMV particles in this manner; both have stimulated the production of neutralizing antibodies in animal studies. The rabies/AlMV chimera developed using this technique was also protective in mice against lethal challenge with the rabies virus.

4.2.1.3   Potato Virus X

Potato virus X (PVX), the type member of the genus Potexvirus, is also a rod-shaped virus containing a monopartite genome. The CP of PVX is expressed from a subgenomic promoter in a similar fashion to the CP of TMV. Unlike TMV, PVX is a flexuous filament particle, and there is no intrinsic size restriction of epitopes or packaging constraints of icosahedral viruses such as CPMV or viruses with tightly packed CP subunits such as TMV (Uhde et al., 2005). As a result, the target epitope can be displayed as many copies on the particle surface (Chapman et al., 1992). However, the structural information for PVX is less complete than it is for TMV and CPMV. Over the past few years, data has accumulated that strongly suggests that the N-terminal portion of the CP is exposed to the surface of the virus particle. As a consequence, this region was thought to be the ideal position in which to insert heterologous peptides into the CP. PVX has been utilized to express VP6 from murine rotavirus, the E7 oncoprotein from human papillomavirus, and a surface antigen from the toxoplasmosis parasite *Toxoplasma gondii*. An HIV epitope of p41 has also been displayed using a PVX antigen presentation system. Mice that were presented with this epitope were shown to produce both IgG and IgA antibodies against HIV.

4.2.2 Polypeptide Expression Systems

4.2.2.1 Tobacco Mosaic Virus

Vaccine genes in their entirety are expressed using the tMV-30B expression vector via an additional copy of the CP subgenomic promoter (see Figure 4.1) (Yusibov et al., 1999). By engineering a leaky termination codon at the C-terminus of the viral CP gene, a TMV-based vector could be developed, resulting in the ability of the modified virus to synthesize both native and recombinant forms of the CP from the same viral RNA (Wigdorovitz et al., 1999). The second promoter is heterogeneous, from a closely related Tobamovirus to reduce the frequency of homologous recombination between promoter elements that may result in the removal of the transgene (Shivprasad et al., 1999; Rabindram and Dawson, 2002). In addition to this, TMV movement and host range through different plant hosts can be improved by small modifications in the nucleotide sequence using a procedure known as *DNA shuffling* (Toth et al., 2002).

α-Trichosanthin was the first therapeutic protein produced in plants using a TMV viral vector. Other full-length proteins produced using the TMV expression system include both single-chain and full-length monoclonal antibodies, as well as VP1 of FMDV, a tumor-derived scFv, a major birch pollen antigen, and the cytosolic form of bovine herpes virus gD protein. Several of these proteins will be described in more detail in Section 4.2.5.1.

4.2.2.2 Potato Virus X

Two types of PVX-based viral vectors have been generated for full-length protein production; one of these involves the duplication of the CP subgenomic promoter in a manner analogous to the TMV-30B system mentioned earlier (Chapman et al., 1992). Inserts as large as GFP can be fused to the N-terminus of the PVX CP and expressed effectively in infected plant cells. Examples of protein expressed using this PVX-duplicated subgenomic promoter-based system include the E7 of human papillomavirus type 16,

VP6 of murine rotavirus, and various human ScFv antibodies (Mallory et al., 2002). The second PVX-based vector that was developed produced a fusion protein consisting of the foreign protein fused to the N-terminus of the PVX CP. The 2A catalytic sequence of FMDV was placed in between these two heterologous sequences and can promote cotranslational cleavage between the foreign insert and the CP of PVX. A low percentage of fusion proteins is incorporated into the virus capsids along with the PVX CP since cleavage is not completely efficient. An ScFv-CP fusion protein was found to be functional when incorporated into PVX virions using this technique. For the sequence of rotavirus VP6, uncleaved VP6-2A-CP was incorporated into PVX virions while the VP6-2A cleavage product formed VLPs (virus-like particles) (see Table 4.1, Figure 4.1 for selected examples). This construct was incompatible with the regulatory requirements that govern production of therapeutic proteins, so Toth et al. (2001) created a PVX vector with an internal ribosome entry site (IRES) between the gene of interest and the resulting vector, producing a bicistronic mRNA that could express both the CP gene and the gene of interest. Further examples of vaccine proteins expressed by polypeptide expression systems are described in the following text.

4.2.2.3   Cowpea Mosaic Virus (CPMV)

CPMV, the type species of the genus Comovirus, has also been used to express full-length proteins. Heterologous proteins are expressed either as coat protein or movement protein fusions with an integral proteolytic cleavage site to allow the target protein to be released (Verver et al., 1998) or as C-terminal fusions with the S protein, incorporating the FMDV 2A peptide as described for PVX (Gopinanth et al., 2000).

4.2.2.4   Plum Pox Virus (PPV)

As in the case of the viruses discussed earlier, PPV (a member of the genus *Potyvirus*) has been developed as an expression vector for whole proteins (Lopez-Moya et al., 2000). An important difference between PPV and the other viruses, however, is that the monopartite genome encodes a single polyprotein from which all the individual virus proteins are released by proteolytic cleavage. The general strategy to express whole proteins using PPV has therefore been to bracket the transgene with recognition sites for the viral protease VPg. There has been only one report of a vaccine candidate expressed in PPV thus far, and that is the VP60 protein from rabbit hemorrhagic disease virus (RHDV), which was inserted between the polymerase and coat protein coding regions (Fernandez-Fernandez et al., 2001). Rabbits immunized with leaf extract containing VP60 demonstrated pathogen-specific immune responses that protected the animals against a lethal challenge with RHDV. No

RHDV was detected in the livers of the surviving animals two weeks after challenge. The serological responses of animals vaccinated with plant extracts containing VP60 were almost as high as those of animals immunized with a commercial vaccine.

4.2.2.5   Cucumber Mosaic Virus (CMV)

CMV is the type species of the genus Cucumovirus and contains a tripartite RNA genome and wide host range. A pseudorecombinant strain of the virus, comprising RNA3 from the CMV-S strain (containing the coat protein gene) and RNAs 1 and 2 from the CMVD strain, was constructed by Natilla et al. (2004). The R9 synthetic epitope of hepatitis C virus E protein (HCV), was introduced into three separate sites in the coat protein gene of CMV. Significant immunoreactivity was detected to crude extracts from plants infected with these chimeric CMV particles and administered to 60 patients suffering chronic HCV (Piazzola et al., 2005). Evidence was also obtained to demonstrate that the chimeric R9-CMV elicits a specific humoral response in rabbits.

4.2.3   PLANT DNA VIRUSES AS EXPRESSION VECTORS

Plant viruses with DNA genomes, such as the Caulimovirus cauliflower mosaic virus (CaMV) and geminiviruses, have been explored as potential protein expression systems as well. CaMV is difficult to work with due to its polycistronic genome, RNA-mediated mode of replication, and restricted host range. On the other hand, geminiviruses possess a simple genomic organization and broad host range, and have been viewed as far more attractive candidates to develop as vectors for foreign gene expression. Geminiviruses can accumulate to extremely high copy numbers in inoculated cells, resulting in greatly elevated levels of gene expression. The geminivirus CP gene can be replaced with a foreign gene, and the resulting recombinant virus displays increases of viral DNA as high as 300,000 copies per cell, indicating that foreign protein expression can be enhanced enormously (Timmermans et al., 1994).

As a novel twist on the classic geminivirus strategy, a mastrevirus known as bean yellow dwarf virus was engineered into an expression vector. The gene encoding the replication initiation protein, or Rep, was placed under independent promoter control, and the SIR and LIR elements, respectively, which are the *cis*-acting elements required for virus replication, were subcloned into another plasmid along with a reporter gene (Hefferon and Dugdale, 2003). This enabled Rep to initiate viral replication and high levels of reporter gene expression. Since Rep can be placed under either inducible or developmental control and initiate viral replication and gene expression when so desired, the implications of these results are great.

The foreign gene of interest can then remain latent within a chromosomal location until initiation of viral replication proceeds when Rep signals the excision of the geminivirus-based *cis*-acting replication machinery (Moon and Hefferon, 2005).

It should be noted that gene silencing is avoided by inserting the geminiviral replicon into stably transformed plants at a chromosomal locus, and the production of proteins that are toxic to the plant can be controlled by regulating promoter expression (e.g., ethanol-inducible, jas-inducible, and seed-specific promoters). By expressing the geminivirus-based vector chromosomally, expression can take place in all plant tissues, and host-species specificity is also avoided. The fact that this strategy will not yield mature viral particles that in turn can cause secondary infection ensures its safeness as a platform for protein expression.

4.2.4 NOVEL STRATEGIES FOR THE DEVELOPMENT OF VIRAL VECTORS

This chapter has so far described the construction of plant viral expression vectors that possess modifications to carry and express heterologous sequences that encode a foreign gene or epitope but behave much like wild-type viruses. These functional viruses have the ability to move systemically through their host, retain infectivity, and produce infectious virus particles. However, problems have been demonstrated to exist with this "full-virus strategy." Large inserts of foreign peptides cannot be tolerated within the virus, as the virus no longer systemically move efficiently or become unstable. Similarly, for virus expression vectors that use an antigen presentation system, only small epitopes no longer than a short span of amino acids can be fused to the CP. In addition to these problems, as the infection process is asynchronous as it progresses at different speeds in different tissues, the host range can be narrow, and virus infection may not reach all harvestable regions of the plant (the lower leaves, for example).

More recently, alternative approaches have been developed that utilize less host-specific plant viral expression vectors as well as improve safety by eliminating the possibility of infectious virus particles that may escape from the host plant from being generated (Marillonnet et al., 2003). One strategy is to place genes that are essential from a virus expression vector into the genome of a transgenic plant. For example, a host plant can express AIMV replicase and thus permit replication of cells that have been infected with a vector containing RNA3 alone, thus enabling replication to take place. This not only simplifies vector development but provides biological containment, as the vector will only be able to replicate in a transgenic line expressing the appropriate viral gene product. In this way, Belanger et al. (2000) used AIMV to display two peptides containing amino acids

174–187 of respiratory syncytial virus (RSV) G-protein. The particles generated strong B- and T-cell responses in primates (Yusibov et al., 2005). The application of this vector expression system is described in more detail later in this chapter.

Using this complementation approach, virus amplification can take place by integration into the host genome or by agrodelivery, followed by controlled release from a chromosomal location. The inability of the expression vector to spread from cell to cell can be compensated in this way by the release of replicons throughout the host from a chromosomally encoded proreplicon or provirus (Mori et al., 2001). Using this "deconstructed virus strategy," a gene of interest can be delivered to the plant via agroinfection (a mode of delivery considered to be more efficient than infection with several assembled viruses) or by site-specific recombination (using recombinases such as Cre or Streptomyces phage c31 integrase, which is highly efficient in plant cells (Kopertekh et al., 2004). The resulting RNA is cleared of residual recombination sites that were previously engineered as portions of introns and is subsequently spliced out by the nuclear RNA-processing machinery.

Using a novel TMV-based vector system designed by Icongenetics, Inc., differentially transformed agrobacteria were used to deliver the viral vector and gene of interest in the form of various modules (Gleba et al., 2004). A library of 5′ modules was developed to target specific plant subcellular organelles (such as the cytosol, rough endoplasmic reticulum, chloroplast, and apoplast). A purification tag (with or without a cleavage sequence) could also be added to the polypeptide of interest. A site-specific recombinase was used to assemble the modules inside the plant cell, and upon transcription, all undesired elements such as recombination sites were removed by splicing events. By combining different elements that contain, for example, targeting or signal peptides, binding domains, protein-encoding fragments, purification domains, or cleavage sites, this approach allows the rapid assembly and expression of arrays of protein variants, and the end result is an extremely high yield of the desired protein.

Using this deconstructed virus strategy, different genes can be cloned into 3′ modules and coinfiltrated along with the same 5′ provector module (Figure 4.2). By deconstructing and reconstructing the viral RNA vector, an enhanced versatility and efficiency is provided because of the numerous gene/vector combinations that can be made and tested without the need to actually make each individual variant construct. The protein under examination can also be targeted to different subcellular compartments or fused to a variety of tags or coding sequence fragments.

The success of this technology largely depended on overcoming a bottleneck in the inability of the vector to form active replicons after delivery

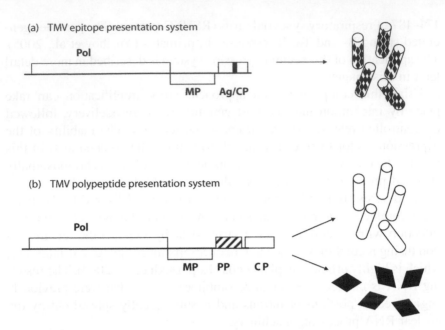

FIGURE 4.2 Schematic diagram of RNA virus expression vectors. (a) TMV as an epitope presentation system, and (b) a polypeptide presentation system. Dark diamonds represent foreign antigen/peptide.

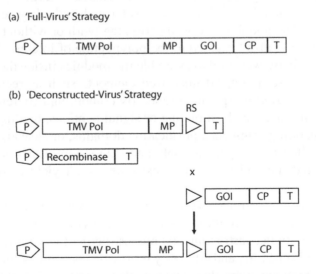

FIGURE 4.3 Schematic representation comparing the full and deconstructed virus vector strategies. (a) TMV expression vector. (b) Provector system for rapid assembly of viral amplicons *in planta*. P; promoter, TMV Pol; TMV polymerase, MP; movement protein, GOI; gene of interest, CP; capsid protein, T; terminator, RS; recombination site. (Adapted from Gleba et al. (2004). *Curr. Opin. Plant Biol.*, 7, 182–188.)

(Marillonnet et al., 2005). Since TMV, an RNA virus, replicates in the cytoplasm rather than the nucleus, it has therefore evolved in an environment where it has not been exposed to the nuclear pre-mRNA-processing machinery. As a result, transcripts of TMV pre-mRNA in the nucleus may be improperly recognized and processed. The authors incorporated silent nucleotide substitutions into the vector and added multiple introns, as well as removing cryptic splice sites (thymine-rich intron-like sequences that could be improperly recognized by the plant cell's RNA-processing machinery) to correct the situation. This correction brought about yields in efficiency as high as 94% of processing of the DNA into active RNA replicons in cells (Marillonnet et al., 2005).

The team from Icongenetics used the deconstructed virus system to attain protein yields of up to 80% of total soluble protein or 5 g/kg of freshwater biomass, a process that only takes 3–14 days. Development of a scaled-up version of this expression system in transgenic plants by controlling the activation of an encrypted version of a replicon present in the plant chromosome of a production host is currently under investigation. One enormous advantage to the system would be the lack of requirement of systemic movement of the virus.

4.2.5 COMMERCIAL USES OF VIRUS EXPRESSION VECTORS

4.2.5.1 Vaccines and Therapeutic Proteins

Several plant viral vectors have been used to successfully produce vaccines and therapeutic proteins in plants. TMV-based expression vectors represent one of the more successful examples and have produced a wide array of therapeutic proteins such as α-trichosanthin, tumor-specific single-chain antibodies, and a number of vaccine antigens (Table 4.1) (Dalsgaard et al., 1997). More recent examples of plant viral expression vectors utilized for vaccine production are provided in detail in Koprowski and Yusibov (2001), Pogue et al. (2002), and Canizares et al. (2005).

Bovine hemorrhagic virus (BHV) is of significant worldwide importance and causes decreased milk production and abortion in pregnant cows. BHV-1 is the causative agent of a group of respiratory and reproductive disorders in cattle. Vaccines against BHV-1 currently in use have been formulated with either inactivated or modified live virus. The inactivated version of the vaccine is a poor immunogen and can cause clinical disease if not inactivated sufficiently. The live vaccine, on the other hand, may induce immunosuppression, so there are disadvantages to both. Unglycosylated forms of the glycoprotein gD were unable to elicit a protective immune response when produced in *E. coli*. Perez-Filgueira et al. (2003) were able to produce a truncated version of the glycoprotein D that

TABLE 4.1

Selected Examples of Vaccines and Therapeutic Proteins Expressed in Plants Using Plant Viral Expression Vectors

A. Fusion Proteins

Virus	Vaccine/Therapeutic Protein	Reference
TMV	HIV-1 peptide	Yusibov et al., 1997
TMV	Malarial peptide	Turpen et al., 1995
PVX	Major birch antigen bet v 1	Wagner et al., 2004
PPV	VP2 peptide from canine parvovirus	Fernandez-Fernandez et al., 1998
CPMV	HIV-1 peptide	Porta et al., 1994
CPMV	Human rhinovirus 14 CP peptide	Porta et al., 1996
CPMV	FMDV VP1 epitope	Wu et al., 2003
CPMV	Canine parvovirus VP2 epitope	Porta et al., 1996
AlMV	Rabies virus	Yusibov et al., 2002
TMV	α-Trichosanthin	Kumagai et al., 1993
TMV	SvFv-CP	McCormick et al., 1999
TMV	BHV-1	Kumagai et al., 1993
TMV	hGH	Gils et al., 2005
TMV	Bovine rotavirus VP8	Perez-Filgueira et al., 2004
BeYDV	SEB	Hefferon and Fan, 2004

resides in the cytosol of plants (gDc) using a genetically engineered version of the TMV expression vector described earlier (TMV-30B). This external membrane protein of the virus has a relatively high molecular weight of 55 kDa. Retention of antigenic integrity of the protein was demonstrated by mechanical inoculation of *Nicotiana benthamiana* tobacco plants with infectious RNA transcripts of the TMV-gDc vector and parental immunization of crude extracts of infected leaves to mice and cattle, which produced both humoral and cellular-specific responses recognizing both the gDc protein as well as BHV-1 virus particles. Mice that were used in this study developed a specific and sustained immune response against both gDc and BHV-1 for up to 80 days postinoculation, a time frame comparable to that registered for mice vaccinated with the conventional immunogen. IgG isotypes for BHV-1 were analyzed in these animals, and indicated that IgG1 and IgG2a were present in predominant levels as well as Ig2b and IgG3 to a lesser extent. Inactivated-virus vaccine, on the other hand, produced only IgG1 to any significant level. Both the emergence of symptoms and detection of virus in nasal secretions of cattle were significantly delayed and shortened in these animals, compared with the control animals. Virus shedding in nasal fluids was significantly reduced, and overall symptoms were much less severe, whereas conventionally vaccinated animals showed variable levels of protection. Since sufficient quantities (20 µg/g fresh tissue) of an animal viral antigen could be produced in crude plant extracts, this demonstrates that plant-produced vaccines offer a safe and inexpensive means by which to immunize animals against viral pathogens.

Previously, the same authors showed that the TMV-30B vector could be used to produce the VP8 fragment of the bovine rotavirus structural protein VP8 in plants (Perez-Filgueira et al., 2004). Affinity chromatography was used to purify this viral fragment in a one-step process in which a small HIS-tag was attached to the carboxyl-terminus of the protein. This purified plant-derived VP8 was capable of eliciting a viral-specific antibody response in vaccinated dams; furthermore, it could also induce passive protection in their offspring. The approaches used in this study once again demonstrate that plant-derived VP8, generated from a TMV-based viral vector, can present a very economical and facile means to produce inexpensive and safe viral antigens in plants.

As an alternative, the TMV expression vector can be used to produce foreign peptides exposed to the virus surface. Expressed epitope peptides are considered to be highly immunogenic because they are exposed at the surface of the virus particle during virus assembly. With a limited length of expressed peptides, a cocktail of TMV vectors expressing several epitopes can be simultaneously produced. TMV coat protein may constitute up to 10% of dry weight of an infected leaf and is easily purified from plant tissue in the form of virus particles. Since the epitopes contain only a small fragment of the antigenic peptide, they present less of a biohazard and thus cannot create any disease. TMV-based expression systems have been employed to present a number of epitopes (Usha et al., 1993; Hamamoto et al., 1993; Fitchen et al., 1995; Turpen et al., 1995; Dalsgaard et al., 1997; Modelska et al., 1998). Recently, TMV has been demonstrated to express epitopes that are highly immunogenic against the foot-and-mouth-disease virus. In a preliminary analysis, parenteral injection of these TMV-based epitopes protected pigs to a significant degree (Wu et al., 2003).

PVX has also been developed as an expression system for the production of antigens for immunogenic purposes. In one study, Bet v 1, the major birch pollen allergen, was inserted behind the subgenomic promoter controlling the transcription of the viral coat protein (Wagner et al., 2004). An ELISA was used to detect the allergen from plant extracts. The allergen was also successfully immunologically characterized further using IgE from different patients. PVX has been used in conjunction with cowpea mosaic virus to express single-chain antibodies (scFv) specific for porcine coronavirus and transmissible gastroenteritis virus (TGEV). Oral administration of plant tissue expressing the protein to 2-day-old piglets demonstrated in vivo protection against challenge with TGEV (Monger et al.,

2006). In addition to this, PVX has been successfully used as a vector for the expression of tuberculosis antigen ESAT-6 in the form of a fusion protein with the FMDV 2A cleavage peptide and the coat protein of PVX (Zelada et al., 2006).

The alfalfa mosaic virus (AlMV) and TMV were examined by Yusibov et al. (1997, 2002) as a combined set of two plant virus-based expression vectors for foreign antigen production. The full-length infectious cDNA clone of AlMV RNA3 (AlMV is a tripartite RNA virus) was used for the first expression vector. An in-frame fusion protein consisting of a peptide against the rabies virus was fused to the AlMV CP gene on RNA3 and assembled into virions. Transgenic tobacco plants expressing RNA1 and RNA2, which encode the replication-associated proteins, were provided in *trans*. The TMV CP was replaced with the rabies epitope peptide sequence AlMVCP fusion protein to create the second expression vector used, based on an autonomously replicating TMV-based vector. Spinach leaves infected with recombinant virus bearing the rabies epitope were fed to mice by Yusibov and coworkers to demonstrate the synthesis of mucosal IgA I. The same group showed in a later study that mice immunized with recombinant virions recovered from infected tobacco plants and were protected against a lethal challenge infection with rabies virus for at least 120 days following parenteral immunization with three doses of recombinant AlMV particles. The authors also found that immunization of mice with the peptide expressed in the context of the plant virus particle made them more immunogenic than with the administration of an equivalent amount of the same peptide in an adjuvant. In addition, they demonstrated that human volunteers fed raw spinach leaves containing experimental plant virus-based rabies vaccine develop a rabies virus-specific immune response.

Geminivirus expression systems can also produce vaccine proteins in plants. A synthetic, plant-optimized version of Staphylococcus endotoxin B (SEB) inserted into a BeYDV reporter cassette displayed high levels of SEB in a plant cell line using ELISA (Hefferon and Fan, 2004). The TMV-based deconstructed virus expression strategy described earlier was employed in the same manner to produce high levels of human growth hormone (hGH) in transfected *N. benthamiana* leaves. The intrinsic flexibility of the system was used to target hGH to the apoplast, chloroplast or cytosol of plants. The highest yield of hGH was found within the apoplast (yield is 10% of total soluble protein or 1 mg/g fresh weight leaf biomass), and the recovered protein was correctly processed and biologically active (Gils et al., 2005).

4.2.5.2   Virus Expression Vectors and Virus-Induced Gene Silencing

Virus-induced gene silencing (VIGS) refers to the silencing of endogenous plant genes using recombinant plant virus vectors (Voinnet et al., 1999; Lindbo et al., 2001; LaComme et al., 2003; Lu et al., 2003; Waterhouse et al., 2001; Waterhouse and Helliwell, 2003; Robertson, 2004). Virus vectors have been utilized to silence endogenous plant genes and require the cloning of homologous gene fragments into the virus without compromising viral replication and movement. The extent to which silencing spreads and the severity of viral symptoms can vary significantly in different host plants and host/virus combinations.

VIGS is powerful in its rapid initiation of silencing in intact wild-type or transgenic plants. The ability of VIGS to silence 1–2 genes reliably can provide material for biochemical analysis, metabolic profiling, and transcript profiling, if suitable controls are included (Burton et al., 2000; Fitzmaurice et al., 2002; Wang and Waterhouse, 2002; Pogue et al., 2002; Helliwell and Waterhouse, 2003). This virus-based vector system approach enables gene fragments to be inserted in sense or antisense orientations into a plant virus expression vector and can lead to systemic knockout of gene expression through posttranscriptional virus-induced gene silencing. Examples of virus-induced silencing by plant viral expression vectors are found in Table 4.2.

One advantage of the virus-based vector system for applications of gene silencing technology is that it is a more rapid procedure than that of generating transgenic plants. It takes 3–4 weeks to follow the procedure from the cDNA stage to the isolation of plants with a silenced phenotype. Still another advantage is that since the overexpression or silencing does not occur until the plant is infected with the viral construct, potentially lethal knockouts can still be expressed.

TABLE 4.2
Selected Examples of Virus-Induced Gene Silencing Using Plant Viral Expression Vectors

Target Gene	Virus	Host	Silencing Phenotype	Reference
NbPDS	TMV	*N. benthamiana*	Photobleaching	Lacomme et al., 2003
NbCesA	PVX	*N. benthamiana*	Dwarf plant	Burton et al., 2000
NbCDPK2	PVX	*N. benthamiana*	Wilting/HR	Romeis et al., 2001
NbCDPK1	PVX	*N. benthmiana*	Cell death	Lee et al., 2003
NbWIPK	PVX	*N. benthamiana*	HR	Sharma et al., 2003

4.2.5.3   Virus Expression Vectors and High-Throughput Gene Function Discovery

Procedures for ultrahigh throughput inoculation of seedlings and proto-plasts, both in bulk and in microtiter well format using viral vector systems, have been developed based upon TMV and bushy stunt mosaic virus (BSMV) vector systems (Fitzmaurice et al., 2002; Holzberg et al., 2002). For example, the TMV-30B recombinant vector containing the dual sub-genomic promoter has been used for several studies and first showed that plant metabolic pathways could be altered to produce novel compounds via epigenetic expression of foreign genes. The TMV vector can be employed as a vehicle for construction of gene libraries and then utilized to infect host plants. By adapting viral vectors via a high-throughput format, novel phenotypes could be observed that may be completely missed using trans-genic approaches. This virus-based vector system can hasten the process of plant functional genomics for crop improvement.

4.3   SUMMARY

This chapter dealt with the more recent developments in the design of plant virus-based expression vectors and their applications in a variety of uses, including vaccine development and gene function discovery. For example, the employment of plant virus capsid proteins as carrier molecules for fused antigenic peptides is attractive as such carrier proteins can self-assemble and form virus particles with the antigenic peptide displayed on their surfaces. Immunogenicity is significantly increased because multiple copies of the foreign antigen can be displayed on the surface of a macromolecular assembly. Modified virus particles can be amplified to a much larger scale and easily purified. Expression of vaccine proteins at high levels using plant virus-based expression vectors will serve as attractive alternatives for a means of generating fast, cost-effective, and safe vaccines for develop-ing countries. Of enormous benefit is the application of plant virus-based vectors for gene function. Many plant virus expression vectors replicate efficiently in specific host plants such as *Nicotiana*. In the case of plant-made biopharmaceuticals, more efforts should be made to develop virus expression systems that infect host plants suitable for oral delivery, such as legumes and cereals. Techniques such as these can be used to provide a myriad of information regarding plant gene function, which can then be applied for many purposes, such as basic research as well as agricultural development. Plant virus-based expression vector technologies have the capacity to provide a wealth of services in both medicine and plant biology for many years to come.

REFERENCES

Belanger, H., Fleysh, N., Cox, S. et al. (2000). Human respiratory syncytial virus vaccine antigen produced in plants. *FASEB J.* 14: 2323–2328.

Bendahmane, M., Koo, M., Karrer, E., and Beachy, R.N. (1999). Display of epitopes on the surface of tobacco mosaic virus: impact of charge and isoelectric point of the epitope on virus-host interactions. *J. Mol. Biol.* 290, 9–20.

Burton, R.A., Gibeaut, D.M., Bacic, A., Findlay, K., Roberts, K. et al. (2000). Virus-induced silencing of a plant cellulose synthase gene. *Plant Cell* 12: 691–706.

Canizares, M.C., Nicholson, L., and Lomonossoff, G.P. (2005). Use of viral vectors for vaccine production in plants. *Immunol. Cell Biol.* 83: 263–270.

Chapman, S., Kavanagh, T., and Baulcombe, D. (1992). Potato virus X as a vector for gene expression in plants. *Plant J.* 2(4): 549–557.

Charterji, A., Burns, L.L., Taylor, S.S. et al. (2002). Cowpea mosaic virus: from the presentation of antigenic peptides to the display of active biomaterials. *Intervirology* 45: 362–370.

Dalsgaard, K., Uttenthal, A., Jones, T.D., Xu, F., Merryweather, A., and Hamilton, W.D.O. (1997). Plant-derived vaccine protects target animals against viral disease. *Nat. Biotechnol.* 15: 248–252.

Fernandez-Fernandez, M.R., Martinez-Torrecuadrada, J.L., Casal, J.I., and Garcia, J.A. (1998). Development of an antigen presentation system based on plum pox potyvirus. *FEBS Lett.* 427: 229–235.

Fernandez-Fernandez, M.R., Mourino, M., Rivera, J. et al. (2001). Protection of rabbits against rabbit hemorrhagic disease virus by immunization with the VP60 protein expressed in plants with a potyvirus-based vector. *Virology* 280: 283–291.

Fitchen, J., Beachy, R.N., and Hein, M.B. (1995). Plant virus expressing hybrid coat protein with added murine epitope elicits autoantibody response. *Vaccine,* 13: 1051–7.

Fitzmaurice, W.P., Holtzberg, S., Lindbo, J.A., Padgett, H.S., Palmer, K.E., Wolfe, G.M., and Pogue, G.P. (2002). Epigenetic modification of plants with systemic RNA viruses. *OMICS* 6(2): 137–151.

Gils, M., Kandzia, R., Marillonnet, S., Kilmyuk, V., and Gleba, Y. (2005). High-yield production of authentic human growth hormone using a plant virus-based expression system. *Plant Biotechnol. J.* 3(6): 613–620.

Gleba, Y., Marillonnet, S., and Klimyuk, V. (2004). Engineering viral expression vectors for plants: the "full virus" and the "deconstructed virus" strategies. *Curr. Opin. Plant Biol.,* 7: 182–188.

Gopinanth, K., Wellink, J., Porta, C. et al. (2000). Engineering cowpea mosaic virus RNA-2 into a vector to express heterologous proteins in plants. *Virology* 267: 159–173.

Hamamoto, H., Sugiyama, Y., and Nakagawa, N. et al. (1993). A new tobacco mosaic virus vector and its use for the systemic production of angiotensin-I-converting enzyme inhibitor in transgenic tobacco and tomato. *Bio/Technology* 11: 930–932.

Hefferon, K.L. and Dugdale, B.G. (2003). Independent expression of Rep and RepA and their roles in regulating bean yellow dwarf virus replication *J. Gen. Virol.* 84: 3465–3472.

Hefferon, K.L. and Fan, Y. (2004). Expression of a vaccine protein in a cell line using a geminivirus-based replicon system. *Vaccine* 23: 404–410.

Helliwell, C. and Waterhouse, P. (2003). Constructs and methods for high-throughput gene silencing in plants. *Methods* 30: 289–295.

Holzberg, S., Brosio, P., Gross, C., and Pogue, G.P. (2002). Barley stripe mosaic virus-induced gene silencing in a monocot plant. *Plant J.* 30: 315–327.

Johnson, J., Lin, T., and Lomonosoff, G. (1997). Presentation of heterologous peptides on plant viruses: genetics, structure and function. *Annu. Rev. Phytopathol.* 35: 67–86.

Kopertekh, L., Juttner, G., and Schiemann, J. (2004). PVX-Cre-mediated marker gene elimination from transgenic plants. *Plant Mol. Biol.* 55(4): 491–500.

Koprowski, H. and Yusibov, V. (2001). The green revolution: plants as heterologous expression vectors. *Vaccine* 19: 2735–2741.

Kumangai, M.H., Turpen, T.H., Weinzenttl, N., della-Cioppa, G., Turpen, A.M., and Donson, J. (1993). Rapid, high-level expression of biologically active alpha-trichosanthin in transfected plants by an RNA viral vector. *Proc. Natl. Acad. Sci. U.S.A.* 90: 427–430.

LaComme, C., Hrubikova, K., and Hein, I. (2003). Enhancement of virus-induced gene silencing through viral-based production of inverted-repeats. *Plant J.* 34: 543–553.

LaComme, C., Pogue, G.P., Wilson, T.M.A., and Santa Cruz, S. (2001). Plant viruses as gene expression vectors. In *Genetically Engineered Viruses*, ed. C.J.A. Ring, E.D. Blair, 59–99. Oxford, UK: BIOS Sci.

Lindbo, J.A., Fitzmaurice, W.P., and della-Cioppa, G. (2001). Virus-mediated reprogramming of gene expression in plants. *Curr. Opin. Plant Biol.* 4, 181–185.

Lomonossoff, G.P. and Hamilton, W.D. (1999). Cowpea mosaic virus-based vaccines. *Curr. Top. Microbiol. Immunol.* 240: 177–189.

Lomonossoff, G.P. and Johnson, G.E. (1996). Use of macromolecular assemblies as expression systems for peptides and synthetic vaccines. *Curr. Opin. Struct. Biol.* 6: 176–182.

Lopez-Moya, J.J., Fernandez-Fernandez, M.R., and Cambra, M. et al. (2000). Biotechnological aspects of plum pox virus. *J. Biotechnol.* 76: 121–136.

Lu, R., Martin-Hernandez, A.M., Peart, J.R., Malcuit, I., Baulcombe, D.C. (2003). Virus-induced gene silencing in plants. *Methods* 30: 296–303.

Mallory, A.C., Parks, G., Endres, M.W., Baulecombe, D., Bowman, L.H., Pruss, G.J., and Vance, V.B. (2002). The amplicon-plus system for high-level expression of transgenes in plants. *Nat. Biotechnol.* 20: 622–625.

Marillonnet, S., Giritch, A., Gils, M., Kandzia, R., Klimyuk, V., and Gleba, Y. (2003). In plant engineering of viral RNA replicons: efficient assembly by recombination of DNA modules delivered by Agrobacterium. *Proc. Natl. Acad. Sci. U.S.A.* 101(18): 6852–6857.

Marillonnet, S., Thoeringer, C., Kandzia, R., Klimyuk, V., and Gleba, Y. (2005). Systemic *Agrobacterium tumefaciens*-mediated transfection of viral replicons for efficient transient expression in plants. *Nat. Biotechnol.* 10: 1038–.

McCormick, A.A., Kumangai, M.H., Hanley, K., Turpen, Y.H., Hakim, I., and Grill, L.K. (1999). Rapid production of specific vaccines for lymphoma by expression of the tumor-derived single-chain Fv epitopes in tobacco plants. *Proc. Natl. Acad. Sci. U.S.A.* 96:703–708.

Modelska, A., Dietzschold, B., Fleysh, N., Fu, Z.F., Steplewski, K., and Hooper, C. (1998). Immunization against rabies with plant-derived antigen. *Proc. Natl. Acad. Sci. U.S.A.* 95: 2481–2485.

Moon, Y.-S. and Hefferon, K.L. (2005). *Geminivirus Replication: Recent Advances in DNA Virus Replication.* Transworld Sciences International.

Monger, W., Alamillo, J.M., Sola, I., Perrin, Y., Bestagno, M., Burrone, O.R., Sabella, P., Plana-Duran, J., Enjuanes, L., Garcia, J.A., and Lomonossoff, G.P. (2006). An antibody derivative expressed from viral vectors passively immunizes pigs against transmissible gastroenteritis virus infection when supplied orally in crude plant extracts. *Plant Biotechnol. J.* 4(6): 623–631.

Mori, M., Fujihara, N., Mise, K., and Furusawa, I. (2001). Inducible high-level mRNA amplification system by viral replicase in transgenic plants. *Plant J.* 27(1): 78–86.

Natilla, A., Piazolla, G., and Nuzzaci, M. et al. (2004). Cucumber mosaic virus as a carrier of a hepatitis C virus-derived epitope. *Arch Virol.* 149: 137–154.

Perez-Filgueira, D.M., Mozgovoj, M., Wigdorovitz, A., Dus Santos, M.J., Parreno, V., Trono, K., Fernandez, F.M., Carrillo, C., Babiuk, L.A., Morris, T. J., and Borca, M.V. (2004). Passive protection to bovine rotavirus (BRV) infection induced by a BRV VP8 produced in plants using a TMV-based vector. *Arch. Virol.* 149: 2337–2348.

Perez-Filgueira, D.M., Zamorano, P.I., Dominguez, M.G., Taboga, O., Del Medico Zajac, M.P., Puntel, M., Romera, S.A., Morris, T.J., Borca, M.V., Sadir, A.M. (2003). Bovine herpes virus gD protein produced in plants using a recombinant tobacco mosaic virus (TMV) vector possesses authentic antigenicity. *Vaccine* 21: 4201–4209.

Piazolla, G., Nuzzaci, M., and Tortorella, C. et al. (2005). Immunogenic properties of a chimeric plant virus expressing hepatitis C virus (HCV)-derived epitope: new prospects for an HCV vaccine. *J. Clin. Immunol.* 25: 142–152.

Pogue, G.P., Lindbo, J.A., Garger, S.J., and Fitzmaurice, W.P. (2002). Making an ally from an enemy: plant virology and the new agriculture. *Annu. Rev. Phytopathol.* 40: 45–74.

Porta, C., Spall, V.E., Findlay, K.C., Gergerich, R.C., Farrance, C.E., and Lomonossoff, G.P. (2003). Cowpea mosaic virus-based chimeras: effects of inserted peptides on the phenotype, host-range and transmissibility of the modified viruses. *Virology* 310: 50–63.

Porta, C., Spall, W.E., Lin, T., Johnson, J.E., Lomonossoff, G.P. (1996). The development of cowpea mosaic virus as a potential source of novel vaccines. *Intervirology* 39: 79–84.

Porta, C., Spall, V.E., Loveland, J., Johnson, J.E., Barker, P.J., and Lomonossoff, G.P. (1994). Development of cowpea mosaic virus as a high-yielding system for the presentation of foreign peptides. *Virology* 202: 949–955.

Rabindram, S. and Dawson, W.O. (2002). Assessment of recombinants that arise from the use of a TMV-based transient expression vector. *Virology* 284: 182–189.

Robertson, D. (2004). VIGS vectors for gene silencing: many targets, many tools. *Annu. Rev. Plant Biol.* 55: 495–519.

Scholthof, H.B., Scholthof, K.-B.G., and Jackson, A.O. (1996). Plant virus gene vectors for transient expression of foreign proteins in plants. *Annu. Rev. Phytopathol.* 34: 299–323.

Shivprasad, S., Pogue, G.P., Lewandowski, D.J., Hidalgo, J., Donson, J., and Grill, L.K. (1999). Heterologous sequences greatly affect foreign gene expression in tobacco mosaic virus-based vectors. *Virology* 255: 312–323.

Timmermans, M.C.P., Das, O.P., and Messing, J. (1994). Geminiviruses and their uses as extrachromosomal replicons. *Annu. Rev. Plant Physiol. Plant. Mol. Biol.* 45: 79–112.

Toth, R.L., Chapman, S., Carr, F., and Santa Cruz, S. (2001). A novel strategy for the expression of foreign genes from plant virus vectors. *FEBS Lett.* 489: 215–219.

Toth, R.L., Pogue, G.P., and Chapman, S. (2002). Improvement of the movement and host range properties of a plant virus vector through DNA shuffling. *Plant J.* 30(5): 593–600.

Turpen, T.H., Reinl, S.J., Charoenvit, Y., Hoffman, S.L., Fallarme, V., and Grill, L.K. (1995). Malarial epitopes expressed on the surface of recombinant tobacco mosaic virus. *Biotechnology* 13: 53–57.

Uhde, K., Fischer, R., and Commandeur, U. (2005). Expression of multiple foreign epitopes presented as synthetic antigens on the surface of potato virus X particles. *Arch. Virol.* 150: 327–340.

Usha, R., Rohl, J.B., Spall, V.E., Shanks, M., Maule, A.J., Johnson, J.E., and Lomonossoff, G.P. (1193). Expression of an animal virus antigenic site on the surface of a plant virus particle. *Virology* 197(1): 366-374.

Verver, J., Wellink, J., and Van Lent, J. et al. (1998). Studies on the movement of cowpea mosaic virus using the jellyfish green fluorescent protection. *Virology* 242: 22–27.

Voinnet, O., Pinto, Y.M., and Baulcombe, D.C. (1999). Suppression of gene silencing: a general strategy used by diverse DNA and RNA viruses of plants. *Proc. Natl. Acad. Sci. U.S.A.* 96: 14147–14152.

Wagner, B., Fuchs, H., Adhami, F., Ma, Y., Scheiner, O., and Breiteneder, H. (2004). Plant virus expression systems for transient production of recombinant allergens in *Nicotiana benthamiana*. *Methods* 32: 227–234.

Wang, M.B. and Waterhouse, P.M. (2002). Application of gene silencing in plants. *Curr. Opin. Plant Biol.* 5: 146–150.

Waterhouse, P.M. and Helliwell, C.A. (2003). Exploring plant genomes by RNA-induced gene silencing. *Nat. Rev. Genet.* 4: 29–38.

Waterhouse, P.M., Wang, M.B., and Lough, T. (2001). Gene silencing as an adaptive defence against viruses. *Nature* 411: 834–842.

Wigdorovitz, A., Perez Filgueira, D.M., Robertson, N., Carrillo, C., Sadir, A.M., and Morris, T.J. (1999). Protection of mice against challenge with foot and mouse disease virus (FMDV) by immunization with foliar extracts from plants infected with recombinant tobacco mosaic virus expressing the FMDV structural protein VP1. *Virology* 255: 312–323.

Wu, X., Dinneny, J.R., Crawford, K.M., Rhee, Y., and Citovsky, V. et al. (2003). Modes of intercellular transcription factor movement in the Arabidopsis apex. *Development* 130: 3735–3745.

Yusibov, V., Hooper, D.C., Spitsin, S.V., Fleysh, N., Kean, R.B., Mikheeva, T., Deka, D., Karasev, A., Cox, S., Randall, J., and Koprowski, H. (2002). Expression in plants and immunogenicity of plant virus-based experimental rabies vaccine. *Vaccine* 20: 3155–3164.

Yusibov, V., Mett, V., and Mett, V. et al. (2005). Peptide-based candidate vaccine against respiratory syncytial virus. *Vaccine* 23: 2261–2265.

Yusibov, V., Modelska, A., Steplewski, K., Agadjanyan, M., Weiner, D., and Hooper, C. (1997). Antigens produced in plants by infection with chimeric plant viruses immunize against rabies virus and HIV-1. *Proc. Natl. Acad. Sci. U.S.A.* 94: 5784–5788.

Yusibov, V., Shivprasad, S., Turpen, T.H., Dawson, W., and Koprowski, H. (1999). Plant viral vectors based on tobacco mosaic viruses. *Curr. Top. Microbiol. Immunol.* 240: 81–94.

Zelada, A.M., Calamante, G., de la Paz Santangelo, M., Bigi, F., Verna, F., Mentaberry, A., and Cataldi, A. (2006). Expression of tuberculosis antigen ESAT-6 in *Nicotiana tabacum* using a potato virus X-based vector. *Tuberculosis (Edinb.)* 86(3–4): 263–267. Epub 2006 Apr. 27.

5 Glycosylation of Therapeutic Proteins in Plants

5.1 INTRODUCTION

Although plants present a promising system for the production of human therapeutic proteins, the majority of such proteins are in fact glycoproteins. In particular, *N*-glycosylation is often essential for the stability, folding, and biological activity of proteins. While transgenic plants possess the intrinsic ability to produce glycoproteins, N-glycoproteins synthesized in plants differ from those derived from their mammalian counterparts. In this chapter, the glycosylation pathways of plants will be described, and the main differences between glycosylation patterns in plant and animal systems will be covered. The potential of animal-based therapeutic proteins produced in plants to behave as allergens, and the modifications between glycoproteins derived from transgenic plants to render them more "humanized," will be discussed. The generation of immunoglobulins, also known as *plantibodies*, and therapeutic glycoproteins, in plants will also be examined.

In the plant cell, newly synthesized proteins are transported to the secretory pathway in the form of "preproteins," where they are cotranslationally inserted into the lumen of the endoplasmic reticulum (ER). These preproteins are targeted by an amino-terminal signal peptide to the ER of the cell. At this stage, upon cleavage of the signal peptide, the proteins are released into the ER lumen (Vitale and Denecke, 1999; Denecke et al., 1990). As they become properly assembled and folded, these proteins can be secreted or targeted to various compartments of the plant cell secretory system (Semenza et al., 1990) [Figure 5.1(a) and (b)]. During this process, many plant proteins undergo a series of complex proteolytic modifications. One of the most significant modifications is glycosylation, which involves the covalent linkage of an oligosaccharide side chain to a protein. For most therapeutic glycoproteins, the oligosaccharide can be attached either to the amide nitrogen of an asparagine residue (known as *N*-glycosylation) and/or the hydroxyl-end of threonine or serine residues in the peptide backbone (known as *O*-glycosylation) (Lerouge et al., 1998; Saint-Jore-Dupas et al., 2007). The general features of these processes will be described in the next section.

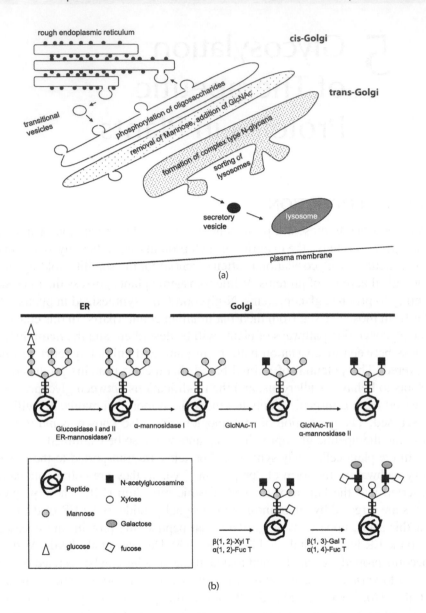

FIGURE 5.1 (a) Movement of glycoproteins through the ER and Golgi. Compartmentalization and addition of glycosyl groups to glycoprotein through Golgi apparatus. Passage of proteins through the plant cell secretory system. Compartments at which proteolytic modifications take place are labeled in red. Movement of proteins through the system are labeled by arrows. (b) *A color version of this figure follows page 110.* Biosynthetic processing of N-glycans in the plant ER and Golgi. Each step indicates the glycosyl groups added to or modified on a nascent polypeptide and the enzymes responsible for this process. (Revised from Chen et al. (2005). *Med. Res. Rev.* 25(3), 343–360.)

5.2 *N*-GLYCOSYLATION

N-linked glycans possess a wide number of roles; they can control glyco-protein folding and conformation; they can protect the glycoprotein from proteolytic degradation; they can present thermal stability, solubility, and biological activity to the glycoprotein; they can provide targeting informa-tion of the glycoprotein to cellular subcompartments or to extracellular secretion; and they can be involved in protein recognition or cell-to-cell adhesion (Lerouge et al., 1998; Chen et al., 2005; Gomord and Faye, 2004; Gomord et al., 2005).

N-glycosylation of plant proteins begins in the ER with the transfer of the oligosaccharide precursor $Glc_3Man_9GlcNAc_2$ from a dolichol lipid carrier to specific asparagine residues on the nascent polypeptide chain. The precursor undergoes a series of successive processing events during transport of the glycoprotein downstream through the secretory pathway as the glycoprotein moves along the ER through the Golgi apparatus to its final destination, often the vacuole (Lerouge et al., 1998). These events involve the removal and addition of sugar residues. During this maturation process, glycosidases and glycosyltransferases located within the ER and Golgi apparatus successively modify the oligosaccharide precursors to a variety of high-mannose-type *N*-glycans ranging from $Man_9GlcNAc_2$ to $Man_5GlcNAc_2$, and then eventually to complex-type *N*-glycans (Lerouge et al., 1998; Chen et al., 2005) [Figure 5.1(a) and (b)].

There is a large heterogeneity of *N*-glycosylation in plants, from the num-ber of glycan side chains, the extent of glycan modification of different side chains on the same glycoprotein and, finally, the amount of heterogeneity of oligosaccharide structures at the same *N*-glycosylation site. This hetero-geneity can be due to the accessibility of processing enzymes of the oli-gosaccharide within the subcellular compartment by processing enzymes performing only partial modifications and by degradation of the glycan side chains by exoglycosidases that may reside within the compartment where the accumulation of the glycoprotein takes place (Chen et al., 2005; Torres et al., 2000; Vitale and Denecke, 1999). For example, Fitchette et al. (1999) examined an *N*-glycan containing a Lewis a (Le[a]) epitope, which plays an important role in cell-to-cell recognition in mammalian cells, is abundant on extracellular glycoproteins, and is never observed on vacuolar glyco-proteins. Le[a] epitopes are both highly abundant and highly immunogenic in mammals. The authors used double-labeling experiments to show that vacuolar glycoproteins do not bypass the late Golgi compartments where Le[a] is synthesized, and the absence of the Le[a] epitope from vacuolar glyco-proteins is most likely the result of its degradation by glycosidases en route to, or after arrival within, the vacuole. By using antiplant Le[a] antibodies as Golgi markers, labeling at these glycoproteins was demonstrated to take place in the trans-Golgi, where the Lewis antigen is likely to be synthesized, suggesting that the formation of the Lewis antigen is a late event in the glycosylation process. Since Le^a-containing N-glycans are only found on extracellular glycoproteins in plants, they can serve as useful markers of the final localization of glycoproteins within the plant cell.

5.3   DIFFERENCES IN N-GLYCOSYLATION PATTERNS BETWEEN PLANTS AND MAMMALS

In contrast to bacteria and yeast, plants are able to produce glycoproteins that possess a core containing two N-acetylglucosamine (GlcNAc) residues, as observed in mammals. The process of modification of glycans in the ER is highly conserved among all species. However, further modification of glycans in the Golgi is highly diverse between plants and mammals. For example, plant N-glycans possess a core $\beta(1,2)$-xylose residue linked to β-mannose and a $\alpha(1,3)$-fucose residue, instead of an $\alpha(1,6)$-fucose residue linked to the proximal glucosamine as in mammals. Complex plant N-glycans are also characterized by the absence of $\beta(1,4)$-galactosyltransferase) and terminal sialic acid (Chen et al., 2005; Gomord and Faye, 2004) (Table 5.1).

5.4   PLANT-MADE PHARMACEUTICALS AND PLANT ALLERGENS

Since posttranslational modifications differ between plants and animals, the question of the immunogenicity of therapeutic glycoproteins in human therapy is raised.

TABLE 5.1
Differences in Glycosylation Patterns between Plants and Mammals

Posttranslational Modification	Mammal	Plant
N-glycosylation	Yes	Yes, minor differences
O-glycosylation	Yes, sugar addition to Ser, Thr only	Yes, sugar addition to Hyp, Ser, Thr
ER retention through signal peptides	Yes	Yes
$\beta(1,2)$-Xylose	No	Yes
$\alpha(1,3)$-Fucose	No	Yes
$\beta(1,4)$-Galactosyltransferase	Yes	No
Sialic acid	Yes	Yes, small amounts
$\alpha(1,6)$-Fucose	Yes	No

For example, the homogeneity of glycosylation patterns of plant-made pharmaceuticals can differ from one development stage to another within a single plant (Elbers et al., 2001). β(1,2)-Xylose and α(1,3)-fucose, which are produced during later stages of glycosylation in plants, are constituents of the glyco-epitopes of some plant allergens. Van Ree et al. (2000) demonstrated that β(1,2)-xylose plays a role in IgE binding; α(1,3)-fucose is also involved. Complex plant *N*-glycans containing β(1,2)-xylose and α(1,3)-fucose are considered to be the major class of so-called carbohydrate cross-reactive determinants that react with IgE antibodies in the sera of many allergic patients.

IgE reactivity to these *N*-glycans have been studied and compared with IgE reactivity to a number of glycoproteins of known primary structures. These differences can induce undesirable immune responses in mammals and/or reduce the activity of recombinant proteins.

Bardor et al. (2003) extracted *N*-glycans from pea, rice, and maize and examined their immunogenicity in rodents. They demonstrated that serum from 50% of nonallergic human blood donors carry antibodies specific for β(1,2)-xylose, and 25% carry antibodies specific for α(1,3)-fucose. These results have great immunological significance; the presence of plant-made pharmaceuticals containing these glycosylation signatures may then induce their rapid immune clearance from the blood stream, thus compromising their effectiveness as therapeutic agents by inducing clinical troubles in allergic populations. In addition to this, hypersensitivity to food allergens can take place in as many as 6%–8% of children. It is possible that sensitization may occur in patients who have prolonged exposure to large quantities of these highly immunogenic *N*-glycans as may be required for certain in vivo therapies. However, a systemically administered recombinant IgG (Guy's 13) synthesized and isolated from plants was demonstrated to result in no immunogenicity in mice despite the differences in glycan groups present in the recombinant antibody (Shillberg, 2003). The results with Guy's 13 antibody imply that plant-derived therapeutic proteins may indeed have useful applications in medicine in spite of the differences observed in their glycosylation patterns.

5.5   STRATEGIES TO HUMANIZE RECOMBINANT PROTEINS IN PLANTS

5.5.1   RETENTION OF GLYCOPROTEINS WITHIN THE ER

Several strategies have been developed to humanize the glycosylation patterns of proteins generated in transgenic plants. The first discussed here is the retention of glycoproteins within the ER.

Many therapeutic proteins, including immunoglobulins and vaccine proteins, are transported from the ER via the Golgi to either the lysosomal compartment, the extracellular matrix, or the blood stream. Most proteins that have been targeted to the ER can be distinguished from cytosolic proteins by an N-terminal signal peptide that associates with a ribonucleoprotein structure known as the signal recognition particle. This signal recognition particle enables cytosolic ribosomes to bind to the ER membrane (Figure 5.2a). Proteins that have been destined for the secretory pathway can then be cotranslationally translocated into the lumen of the ER. Once inside the ER lumen, a cleavage event removes the signal peptide from the polypeptide. The polypeptide can then associate with ER-resident enzymes and molecular chaperones that help to fold it into its native conformation. This process is conserved among all eukaryotes.

The ER is the subcellular compartment in which recombinant proteins have the greatest stability in the plant secretory pathway. Once retained within the ER, newly synthesized proteins will not be exposed to the posttranslational modifications (e.g., the generation of $\beta(1,2)$-xylose and $\alpha(1,3)$-fucose-containing glycan side chains) that would equip them with the potential to be allergens. Reticuloplasmins, or ER-resident proteins, contain a C-terminal retrieval signal consisting of the consensus tetrapeptide K/HDEL (Asp-x-Ser/Thr). This signal enables the retrieval of any proteins that escape to the Golgi apparatus back into the ER via the KDEL receptor located in the cis-Golgi (Vitale and Denecke, 1999; Pagny et al., 2000; Semenza et al., 1990). Proteins that possess the ER retention signal can then be collected from the Golgi and recycled to the ER via transitional vesicles (Figure 5.2b). It should be noted that signal peptides function interchangeably in animal and plant cells; there is no evidence that one or the other system has increased efficiency.

In addition, cytosolic proteins can be modified by the addition of signal peptides and targeted to the lumen of the ER (Denecke et al., 1990). The first cytoplasmic protein targeted to the ER by the addition of a signal peptide was patatin (Denecke et al., 1990). However, further studies have revealed that recombinant glycoproteins fused with the retention sequence HDEL may still exhibit N-glycan modifications that occur in the Golgi apparatus, indicating that such artificial reticuloplasmins can pass through portions of the Golgi network where they undergo further posttranslational modification, then be retrieved and recycled to the ER through a retrograde transport pathway (Pagny et al., 2000). Torres et al. (2001) expressed a single-chain antibody fragment with a KDEL signal and the native molecular chaperone calreticulin, a protein that naturally resides within the ER, to show that both artificial and native reticuloplasmins have identical distribution profiles within the ER network of plant tissue. In addition to this, higher expression levels of therapeutic proteins

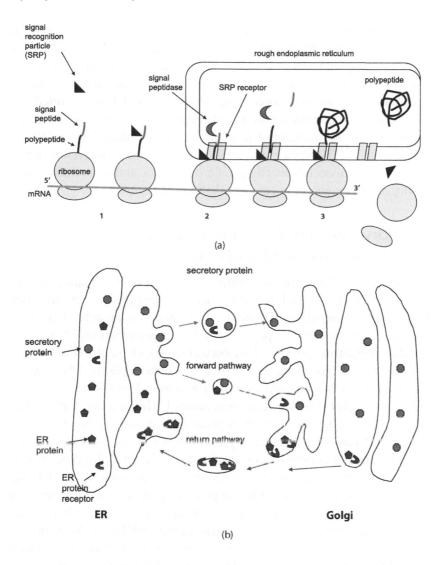

(a)

(b)

FIGURE 5.2 (a) Processing of proteins within the ER. Cleavage of signal peptide and translocation of protein to the ER lumen. (1) During translation, the signal peptide is synthesized first and becomes associated with the signal recognition particle (SRP). (2) The ribosome SRP complex then binds to the SRP receptor. The signal sequence, which is now inserted within the lumen, is cleaved by a signal peptidase. (3) The newly synthesized polypeptide is released into the ER lumen, and the ribosome and SRP complex dissociates. Further details are provided in the text. (b) Retention of ER proteins via the retrograde transport pathway. Receptor proteins located on the membrane are circulated throughout the Golgi network, and they enable ER proteins that contain the ER retention signal (KDEL) to be collected and returned to the ER from the Golgi through transitional vesicles.

have been achieved when the protein of interest was targeted to the ER lumen. For example, Richter et al. (2000) demonstrated that an ER signal peptide from soybean gene vspA or an ER retention signal KDEL was able to influence accumulation of vaccine protein HBsAg in transgenic potato. Increased HBsAg accumulation using a retention signal also resulted in a higher amount of oligomerization of the antigen into virus-like particles, resulting in an improvement of immunogenicity of plant-derived HBsAg (Sojikul et al., 2003). Other strategies are also being employed to target plant-made pharmaceuticals to specific organelles and to maintain their stability in a plant environment (Benchabane et al., 2008).

5.5.2   Development of Plants that Lack β(1,2)-Xylose and Core α(1,3)-Fucose

As mentioned earlier, glycosyl groups that are unique to plants, such as β(1,2)-xylose and core α(1,3)-fucose, are also responsible for increased allergenicity of plant-derived proteins. A number of approaches have been employed to circumvent this problem, and a few are mentioned here.

To date, mutant and gene knocking out technologies have often been used to avoid plant-specific modification of N-glycans. For example, a *cgl Arabidopsis* mutant that lacks N-acetyl glucosaminyltransferase I and is thus unable to synthesize complex type N-glycans was isolated. This was accomplished by screening a large population of mutant plants with an antisera directed toward complex N-glycans found on plant glycoproteins (Von Schaewen et al., 1993). Most of the antibodies in this serum are directed toward β-(1,2)-xylose. Mutant plants whose extracts were nonreactive to the *cgl* serum on immuno-dot blots were determined to be blocked in the processing of N-linked glycans. The plants grew well under optimal growth conditions, suggesting that β(1,2)-xylose and α(1,3)-fucose are not essential for the viability of the plant. The authors crossed the *cgl* mutant with another transgenic *Arabidopsis* and were able to obtain double-transformants that expressed the protein of interest and lacked β(1,3)-fucose and α(1,2)-xylose residues.

Another *Arabidopsis* mutant, *mur1*, which lacks the ability to synthesize l-fucose, possesses a defective gene encoding GDP-d-Man-4,6-dehydratase, a key enzyme in l-fucose biosynthesis. Further analysis revealed that l-Fuc is replaced by l-Gal, a structurally similar monosaccharide, in the cell walls of this mutant with no adverse effects on plant physiology or metabolism (Rayon et al., 1999). Transgenic plants containing this mutation can also be used for foreign protein production.

In 2004, Strasser et al. generated a *xylT* knockout mutant that no longer produced β(1,2)-linked xylosyltransferase, and a *fucTA and fucTB* double knockout mutant that was unable to produce α(1,3)-fucosyltransferase, in

Arabidopsis plants. Alterations in the *N*-glycosylation patterns between wild-type and *fucT* mutant lines were determined by MALDI-TOF mass spectrometry. Absence of xylose residues in the *N*-glycans can be monitored by a reduction of mass of the respective peaks. The resulting plants that were selected can produce *N*-glycans that lack either β(1,2)-linked xylose or α(1,3)-linked fucose. The authors then crossed these two mutant lines to produce β(1,2)-xylose- and core α(1,3)-fucose-deficient plants. Progeny were screened by dot blot hybridization using anti-horseradish peroxidase antibodies, which recognize β(1,2)-xylose- and core α(1,3)-fucose-containing epitopes. These knockout mutant lines were viable, and experienced no deleterious impact on plant growth and development. In a different study, α(1,3)-fucosyltransferase and β(1,2)-xylosyltransferase genes were knocked out in *Physcomitrella patens*. The authors found that they could prevent the production of plant-specific glyco-epitopes without affecting protein secretion in these plants.

Another way that β(1,2)-linked xylose can be removed from plant proteins involves the use of β(1,2)-xylosidase. This is a degradative enzyme that is easily prepared from potatoes. If the 3-position of this mannose is not occupied, β(1,2)-xylosidase releases xylose residues that are β(1,2)-linked to the beta-mannose of an *N*-glycan core. This technique can also be applied to plant-derived therapeutic glycoproteins to remove potential immunogenic epitopes (Lerouge et al., 1998).

In summary, the transgenic plant studies described have indicated that it is possible to knock out genes related to β(1,2)-linked xylose and α(1,3)-linked fucose biosynthesis to produce more mammalian-like glycoproteins (Von Schorwen et al., 1993; Rayon et al., 1999).

5.5.3 EXPRESSION OF HUMAN β(1,4)-GALACTOSYLTRANSFERASE IN PLANTS

The expression of mammalian β(1,4)-galactosyltransferases in plants to complete and/or compete with endogenous machinery for *N*-glycan maturation in the plant Golgi apparatus is another useful strategy in the effort to humanize plant *N*-glycans. β(1,4)-galactosyltransferase (GalT) is the first glycosyltransferase in mammalian cells that initiates the further branching of complex *N*-linked glycans after GnT-I and GnT-II (it transfers Gal from UDP-Gal in β(1,4)-linkage to GlcNAc residues in *N*-linked glycans). The addition of terminal β(1,4)-galactose residues is generally believed to contribute to correct antibody folding and immunological function. Bakker et al. (2001) demonstrated that human GalT can be successfully expressed in tobacco plants and is able to partially humanize *N*-linked glycans of endogenous glycoproteins, as well as transgenically expressed mammalian antibodies. This feature is inheritable, and no phenotypic differences were found between transgenic plants expressing GalT and wild-type plants. The authors expressed a plantibody with 30% galactosylated N-glycans, which was approximately as abundant in N-glycan concentration as that produced by hybridoma cells, by crossing a tobacco plant expressing GalT with a plant expressing heavy and light chains of mouse monoclonal antibody Mgr-48 IgG_1. The results of this study indicated that those N-glycans that are produced in plants possessed much greater variability than those produced in hybridoma cells. Most of this diversity arises from the variability in $\beta(1,4)$-xylose and $\alpha(1,3)$-fucose content. The authors conclude that expression of GalT in tobacco plants results in galactosylation of antibodies similar to that observed in mammals, and that human GalT localizes correctly in the plant Golgi apparatus.

In another study, a tobacco BY2 cell suspension that expressed the human GalT gene, known as the GT6 cell line, was developed (Palacpac et al., 1999). A structural analysis was performed on oligosaccharide moieties from glycoproteins of these transformed cells to demonstrate any changes in plant N-glycan structure and to confirm that galactosylation in fact improved the N-linked pathway in tobacco BY2 cells. Unlike the transgenic tobacco plants analyzed by Bakker et al. (2001), no $\alpha(1,3)$-fucosylated and a total of only 6.6% $\beta(1,2)$-xylosylated glycans could be detected. In addition, while the N-glycan structures of the glycoproteins produced in BY2 cells and secreted to the spent medium were different from the N-glycan structures of intracellular glycoproteins, the glycoproteins of GT6 cells possessed both N-linked glycan structures of extracellular glycoproteins as well as intracellular glycoproteins. The results of this study indicate that these GT6 cells can be used for modification of transgene glycoproteins to a glycosylation signature that is constant and predictable. GT6 cells can also be used to compare galactosylated and nongalactosylated versions of recombinant proteins, thus enabling an assessment of the functional consequences of galactosylation (Palacpac et al., 1999; Misaki et al., 2003).

The efficiency of heterologous glycosyltransferases could be increased by controlling their targeting ability within the Golgi apparatus. The targeted expression of heterologous glycosyltransferases using the fusion of a sequence responsible for targeting of *A. thaliana* and $\beta(1,4)$-xylosyltransferase in the Golgi and the catalytic domain of the human $\beta(1,4)$-galactosyl transferase has been patented (Gomord and Faye, 2004).

5.6 *O*-GLYCOSYLATION

O-glycosylation in plants occurs mainly on the hydroxyl groups of hydroxyproline, serine and threonine residues within Hyp-rich glycoproteins (HGRPs). HRGPs are located either in the plant cell wall or at the

outer surface of the plasma membrane; they represent major surface glycoproteins in plants. O-glycosylation has also been shown to be important for protein functionality. To date, little attention has been paid to
the O-glycosylation status of therapeutic proteins produced in transgenic
plants. It is impossible to predict whether a therapeutic protein of mammalian origin, containing glycans that are O-linked to Thr/Ser residues, will
be correctly O-glycosylated when produced in a plant expression system
(Chen et al., 2005; Gomord and Faye, 2004).

5.7   SIALIC ACID IN PLANTS

Sialylation affects the biological activity and circulatory half-life of many
therapeutically important glycoproteins, including hormones, enzymes,
immunoglobulins, blood factors, and cytokines (Chen et al., 2005). In the
past, it was believed that plants do not perform sialylation of glycoconjugates. Recently, Shah et al. (2003) showed that sialylated glycoconjugates are indeed present at low quantities in suspension-cultured cells of
Arabidopsis thaliana, as well as *Nicotiana tabacum* and *Medicago sativa*.
The authors conclude that specific metabolic engineering of this endogenous plant sialylation pathway may be a more effective means to enhance
the value of plants as production systems for mammalian proteins. More
recently, Paccalet et al. (2007) engineered a sialic acid synthesis pathway
in transgenic plants using enzymes derived from *E. coli* and *C. jejuni*.

5.8   GLYCOSYLATION OF IMMUNOGLOBULINS
PRODUCED IN PLANTS

Immunoglobulins can be utilized as model glycoproteins to determine the
potential of a plant expression system to produce humanized glycoproteins.
Many of the properties of immunoglobulins are dependent upon their
glycosylation patterns. For IgG, for example, there is one conserved
N-glycosylation site in the C_H2 domain per heavy chain of the Fc region.
N-glycosylation of this site is critical to the structural stability of the immunoglobulin molecule. The Fab region is also glycosylated in about 30% of
serum antibodies. A nonglycosylated IgG may exhibit a greater sensitivity
to proteases and a loss of binding capacity to monocyte Fc receptors. The
result of this is that N-glycosylation is a key step for the production of fully
functional immunoglobulins using a heterologous expression system (Ko
and Koprowski, 2005).

The first comparative study of a mammalian glycoprotein produced
in a transgenic plant system was performed by Cabanes-Macheteau et al.
(1999). Guy's 13 antibody, which is specific for a cell-surface protein of

Streptococcus mutans, bacteria that are the principal cause of dental caries in humans, was used in this study. This monoclonal antibody (MAb) contains two potential *N*-glycosylation sites on its constitutive heavy chain, one within the Fc region, and the other within the Fab region of the molecule. Full-length MAb Guy's 13 expressed in tobacco was found to be functional in terms of antigen recognition and binding. A detailed structural comparison of the *N*-glycans of this MAb expressed in mouse and in tobacco plants revealed that the diversity of oligosaccharide structures on the plantibody molecule represents an array of structurally related oligosaccharrides from high-mannose-type *N*-glycans (40%) to modified glycans (60%; Figure 5.3). The heterogeneity of the carbohydrate moiety in the antibody is higher in transgenic tobacco than in mouse. The plantibody contained different *N*-linked oligosaccharides, including β(1,2)-xylose and α(1,3)-fucose residues, which, as mentioned earlier, are known to be highly immunogenic in animals and could be involved in allergic reactions. This study illustrates the differences in the glycosylation pattern that takes place within the same protein backbone; these differences imply that the number of recombinant mammalian glycoproteins produced in transgenic plants may be limited.

In another study, Ko et al. (2003) expressed human antirabies MAb in transgenic tobacco plants and compared its *N*-glycosylation patterns

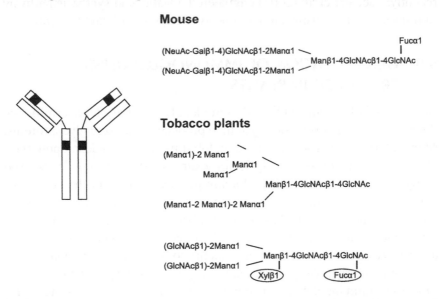

Mouse

(NeuAc-Galβ1-4)GlcNAcβ1-2Manα1
(NeuAc-Galβ1-4)GlcNAcβ1-2Manα1 — Manβ1-4GlcNAcβ1-4GlcNAc — Fucα1

Tobacco plants

(Manα1)-2 Manα1
Manα1
Manα1 — Manβ1-4GlcNAcβ1-4GlcNAc
(Manα1-2 Manα1)-2 Manα1

(GlcNAcβ1)-2Manα1
(GlcNAcβ1)-2Manα1 — Manβ1-4GlcNAcβ1-4GlcNAc — Xylβ1 — Fucα1

FIGURE 5.3 *N*-glycosylation of Guy's 13 monoclonal antibody (MAb) expressed both in mouse and in transgenic tobacco plants. Depicts the site of *N*-glycosylation. Glycan residues within ovals play a role in the immunogenicity of plant *N*-glycans. (Revised from Lerouge et al. (2000). *Curr. Pharm. Biotechnol.* 1: 347–354.)

with both human MAb SO57 expressed in human/murine hybridoma cell lines as well as commercial HRIG. The authors found that the plant-derived human MAb has a neutralizing activity that is comparable to its mammalian-derived counterparts and, in addition, an efficacy in rabies postexposure prophylaxis comparable to HRIG. However, modifications in N-glycosylation patterns did exist. The antirabies MAbs produced in plants also possessed a KDEL signal peptide fused to the heavy chain of the immunoglobulin. The mouse antirabies MAb contained 17 complex N-glycans in the conserved glycoslyation site on each heavy chain, whereas its plant-derived counterpart predictably contained mostly oligomannose-type N-glycans, since it was retained in the ER and was unable to pass further along the secretory pathway than the *cis*-Golgi stack. As a result, no plant-specific $\alpha(1,3)$-fucose or $\beta(1,2)$-xylose residues were found on the plant-derived MAb. Similarly, a comparison of mouse- or plant-derived MAbs that lacked the KDEL sequence possessed a few differences in oligomannose glycan patterns. The altered glycan structures did not appear to affect in any way the specific binding activity of plant-derived MAbs to colorectal cancer cells, for example (Tekoah et al., 2004).

A study conducted by Elbers et al. (2001) examined both soluble endogenous glycoproteins as well as recombinant mouse antibodies from transgenic tobacco leaves to determine whether the adaptation of plant cells to environmental changes can be reflected in their N-glycosylation patterns. This study is important for the production of therapeutic glycoproteins in plants in order to assess the consistency of the quality of glycoproteins. The results of this study indicated that senescence of leaves can indeed influence glycosylation. Antibodies isolated from young, uppermost leaves of the plant have a relatively high amount of high-mannose glycans as compared with antibodies extracted from older leaves from the base of the plant; these older leaves contain more terminal GlcNAc residues. These results suggest that the high-mannose antibodies are gradually processed to terminal GlcNAc-rich complex glycans during leaf maturation. In terms of producing plantibodies, the authors conclude that the age of the leaves from which the antibodies are harvested is an important factor in producing antibodies that maintain a consistent glycosylation signature.

5.9   OTHER THERAPEUTIC PROTEINS

In addition to antibodies, numerous human blood glycoproteins have been produced in transgenic plants. Most are indistinguishable from their mammalian counterparts in terms of amino acid sequence, conformation, and biological activity. However, structural differences can be observed with regard to the electrophoretic mobilities between mammalian glycoproteins and their counterparts in transgenic plants, as well as by affinity- and immunoblots of purified material (Lerouge et al., 2000).

5.10 CONCLUSIONS

While both plant and mammalian-derived proteins possess N-linked and O-linked glycan structures, some differences exist concerning the modification of glycans in the Golgi apparatus. Some of these differences may lead to an increase in allergenicity and an undesirable immune response in mammals that are administered plant-derived glycoproteins. Efforts have been made to further "humanize" plant-derived therapeutic proteins and immunoglobulins by altering a number of glycosylation pathways in plants. Retention of proteins within the ER, designing fucosyltransferase (FucT) and xylosyltransferase (XylT) knockout mutant plants, and the addition of mammalian $\beta(1,4)$-galactosyltransferase to plants are steps in the right direction. Furthermore, the production of biologically active immunoglobulins in plants has indicated that correct folding and assembly can take place in planta, and opens a new venue for which antibodies can be produced in large scale and at low cost (Liénard et al., 2007). Future work, such as engineering plants to express glycoproteins that are correctly sialylated and O-glycosylated, will further enhance the development and applications of plant-derived proteins in medicine.

REFERENCES

Bakker, H., Bardor, M., Molthoff, J.W., Gomord, V., Elbers, I., Stevens, L., Jordi, W., Lommen, A., Faye, L., Lerouge, P., and Bosch, D. (2001). Galactose-extended glycans of antibodies produced by transgenic plants. *Proc. Natl. Acad. Sci. U.S.A.* 98(5): 2899–2904.

Bardor, M., Faveeuw, C., Fitchettew, A.-C., Gilbert, D., Galas, L., Trottein, F., Faye, L., and Lerouge, P. (2003). Immunoreactivity in mammals of two typical plant glyco-epitopes, core $\alpha(1, 3)$-fucose and core $\beta(1, 2)$-xylose. *Glycobiology* 13(6): 427–434.

Benchabane, M., Goulet, C., Rivard, D., Faye, L., Gomord, V., and Michaud, D. (2008). Preventing unintended proteolysis in plant protein biofactories. *Plant Biotechnol. J.* 6(7): 633-648.

Cabanes-Macheteau, M., Fitchette-Laine, A.-C., Loutelier-Bourhis, C., Lange, C., Vine, N.D., Ma, J.K.C., Lerouge, P., and Faye, L. (1999). N-glycosylation of a mouse IgG expressed in transgenic tobacco plants. *Glycobiology* 9(4): 365–372.

Capell, T., Claparois, L., Del Duca, S., Bassie, L., Miro, B., Rodriguez-Montesinos, J., Christou, P., and Serafini-Fracassini, D. (2004). Producing transglutaminases by molecular farming in plants: minireview article. *Amino Acids* 26: 419–423.

Chargelegue, D., Vine, N.D., van Dolleweerd, C.J., Drake, P.M., and Ma, J.K. (2000). A murine monoclonal antibody produced in transgenic plants with plant-specific glycans is not immunogenic in mice. *Transgenic Res.* 9: 187–194.

Chen, Y., Vandenbussche, F., Rouge, P., Poroost, P., Zeumans, W.J., and Van Damme, E.J.M. (2002). A complex fruit-specific type-2 ribosome-inactivating protein from elderberry (*Sambucus nigra*) is currently processed and assembled in transgenic tobacco plants. *Eur. J. Biochem.* 269: 2897–2906.

Chen, M., Liu, X., Wang, Z., Song, J., Qi, Q., and Wang, P.G. (2005). Modification of plant *N*-glycans processing: the future of producing therapeutic protein by transgenic plants. *Med. Res. Rev.* 25(3): 343–360.

Daniell, H., Streatfield, S.J., and Wycoff, K. (2001). Medical molecular farming: production of antibodies, biopharmaceuticals and edible vaccines in plants. *Trends Plant Sci.* 6(5): 219–226.

Denecke, J., Botterman, J., and Deblaere, R. (1990). Protein secretion in plant cells can occur via a default pathway. *Plant Cell* 2, 51–59.

Elbers, I.J.W., Stoopen, G.M., Bakker, H., Stevens, L.H., Bardor, M., Molthoff, J.W., Jordi, W.J.R.M., Bosch, D., and Lommen, A. (2001). Influence of growth conditions and developmental stage on *N*-glycan heterogeneity of transgenic immunoglobulin G and endogenous proteins in tobacco leaves. *Plant Physiol.* 126: 1314–1322.

Fitchette, A.,-C, Cabanes-Macheteau, M., Marvin, L., Martin, B., Stiat-Jeunemaitre, B., Gomord, V., Crooks, K., Lerouge, P., Faye, L., and Hawes, C. (1999). Biosynthesis and immunolocalization of Lewis a-containing *N*-glycans in the plant cell. *Plant Physiol.* 121: 333–343.

Frigerio, L., Vine, N.D., Pedrazzini, E., Hein, M.B., Wang, F., Ma, J.K.-C, and Vitale, A. (2000). Assembly, secretion and vacuolar delivery of a hybrid immunoglobulin in plants. *Plant Physiol.* 123: 1483–1493.

Galeffi, P., Lombardi, A., Di Donato, M., Latini, A., Sperandei, M., Cantale, C., and Giacomini, P. (2005). Expression of single-chain antibodies in transgenic plants. *Vaccine* 23: 1823–1827.

Gomord, V. and Faye, L. (2004). Posttranslational modification of therapeutic proteins in plants. *Curr. Opin. Plant Biol.* 7: 171–181.

Gomord, V., Chamberlain, P., Jefferis, R., and Faye, L. (2005). Biopharmaceutical production in plants: problems, solutions, and opportunities. *Trends Biotechnol.* 24(4): 147-149.

Joshi, L. and Lopez, L. (2005). Bioprospecting in plants for engineered proteins. *Curr. Opin. Biol.* 8: 223-226.

Ko, K. and Koprowski, H. (2005). Plant biopharming of monoclonal antibodies. *Virus Res.* 111: 93–100.

Ko, K., Techosh, Y., Rudd, P.M., Harvey, D.J., Dwek, R.A., Spitsin, S., Hanlon, C.A., Rupprecht, C., Dietzschold, B., Golovkin, M., and Koprowski, H. (2003). Function and glycosylation of plant-derived antiviral monoclonal antibody. *Proc. Natl. Acad. Sci. U.S.A.* 100(13): 8013–8018.

Larrick, J.W., Yu, L., Naftzger, C., Jaiswal, S., and Wycoff, K. (2001). Production of secretory antibodies in plants. *Biomol. Eng.* 18: 87–94.

Lerouge, P., Cabanes-Macheteau, M., Rayon, C., Fischette-Laine, A.-C., Gomord, V., and Faye, L. (1998). *N*-glycoprotein biosynthesis in plants: recent developments and future trends. *Plant Mol. Biol.* 3: 31–48.

Lerouge, P., Bardor, M., Pagney, S., Gomord, V., and Faye, L. (2000). *N*-glycosylation of recombinant pharmaceutical glycoproteins produced in transgenic plants: towards a humanisation of plant *N*-glycans. *Curr. Pharm. Biotechnol.* 1: 347–354.

Liénard, D., Sourrouille, C., Gomord, V., and Faye, L. (2007). Pharming and transgenic plants. *Biotechnol. Annu. Rev.* 13: 115-147

Ma, J.K.-C. and Hein, M. (1995). Immunotherapeutic potential of antibodies produced in plants. *Trends Biotechnol.* 13: 522–527.

Ma, J.K.-C., Drake, P.M., Chargelgue, D., Obregon, P., and Prada, A. (2005). Antibody processing and engineering in plants, and new strategies for vaccine production. *Vaccine* 23: 1814–1818.

Misaki, R., Kimura, Y., Palacpac, N.Q., Yoshida, S., Fujiyama, K., and Seki, T. (2003). Plant cultured cells expressing human β1,4-galactosyltransferase secrete glycoproteins with galactose-extended *N*-linked glycans. *Glycobiology* 13(3): 199–205.

Mokrzycki-Issartel, N., Bouchon, B., Farrat, S., Berland, P., Laparra, H., Madelmont, J.-C., and Theisen, M. (2003). A transient tobacco expression system coupled to MALDI-TOF-MS allows validation of the impact of differential targeting on structure and activity of a recombinant therapeutic glycoprotein produced in plants. *FEBS Lett.* 552: 170–176.

Paccalet, T. et al. (2007). Engineering of a sialic acid synthesis pathway in transgenic plants by expression of bacterial Neu5AC-synthesizing enzymes. *Plant Biotechnol. J.* 5(1): 16-25.

Pagny, S., Cabanes-Macheteau, M., Gilikin, J.W., Leborgne-Castel, N., Lerouge, P., Boston, R.S., Faye, L., and Gomord, V. (2000). Protein recycling from the Golgi apparatus to the endoplasmic reticulum in plants and its minor contribution to calreticulin retention. *Plant Cell* 12: 739–755.

Palacpac, N.Q., Yoshida, S., Sakai, H., Kimura, Y., Fujiyama, K., Yoshida, T., and Seki, T. (1999). Stable expression of human β1,4-galactosyltransferase in plant cells modifies *N*-linked glycosylation patterns. *Proc. Natl. Acad. Sci. U.S.A.* 96: 4692–4697.

Rayon, C., Cabanes-Macheteau, M., Loutelier-Bourhis, C., Salliot-Marie, I., Lemoine, J., Reiter, W.-D., Lerouge, P., and Faye, L. (1999). Characterization of *N*-glycans from *Arabidopsis*. Application to a fucose-deficient mutant. *Plant Physiol.* 119: 725–733.

Richter, L.J., Thanavala, Y., Arntzen, C.J., and Mason, H.S. (2000). Production of hepatitis B surface antigen in transgenic plants for oral immunization. *Nat. Biotechnol.* 18: 1167–1171.

Saint-Jore-Dupas, C., Faye, L., and Gomord, V. (2007). From planta to pharma with glycosylation in the toolbox. *Trends Biotechnol.* 25(7): 317-323.

Schillberg, S., Fischer, R., and Emans, N. (2003). "Molecular farming" of antibodies in plants. *Naturwissenschaften* 90: 145–155.

Semenza, J.C., Hardwick, K.G., Dean, N., and Pelham, H.R. (1990). ERD2, a yeast gene required for the receptor-mediated retrieval of luminal ER proteins from the secretory pathway. *Cell* 61: 1349–1357.

Shah, M.M., Fojiyama, K., Flynn, C.R., and Joshi, L. (2003). Sialylated endogenous glycoconjugates in plant cells. *Nat. Biotechnol.* 21(12): 1470–1471.

Sojikul, P., Buehner, N., and Mason, H.S. (2003). A plant signal peptide-hepatitis B surface antigen fusion protein with enhanced stability and immunogenicity expressed in plant cells. *Proc. Natl. Acad. Sci. U.S.A.* 100(5): 2209–2214.

Stoger, E., Sack, M., Tischer, R., and Christou, P. (2002). Plantibodies: applications, advantages and bottlenecks. *Curr. Opin. Biotechnol.* 13: 161–166.

Strasser, R., Altmann, F., Mach, L., Glossl, J., and Steinkellner, H. (2004). Generation of *Arabidopsis thaliana* plants with complex N-glycans lacking β1,2-linked xylose and core α1,3-linked fucose. *FEBS Lett.* 561: 132–136.

Teloah, Y., Ko, K., Koprowski, H., Harvey, D.J., Wormald, M.R., Dwek, R.A., and Rudd, P.M. (2004). Controlled glycosylation of therapeutic antibodies in plants. *Arch. Biochem. Biophys.* 426: 266–278.

Torres, E., Gonzalez-Melendi, P., Stuger, E., Shaw, P., Twyman, R.M., Nicholson, L., Vaquero, C., Fischer, R., Christou, P., and Perrin, Y. (2001). Native and artificial reticuloplasmins co-accumulate in distinct domains of the endoplasmic reticulum and in post-endoplasmic reticulum compartments. *Plant Physiol.* 127: 1212–1223.

Van Ree, R., Cabanes-Macheteau, M., Akkerdaas, J., Milazzo, J.-P., Loutelier-Bourhis, C., Rayon, C., Villalba, M., Koppelman, S., Aalberse, R., Rodriguez, R., Faye, L., and Lerouge, P. (2000). β(1,2)-xylose and α(1,3)-fucose residues have a strong contribution in IgE binding to plant glycoallergens. *J. Biol. Chem.* 273(15): 11451–11458.

Vitale, A. and Denecke, J. (1999). The endoplasmic reticulum: gateway of the secretory pathway. *Plant Cell* 11: 615–628.

Von Schaewen, A., Sturm, A., O'Neill, J., and Chripeels, M. (1993). Isolation of a mutant *Arabidopsis* plant that lacks N-acetyl glucosaminyl transferase I and is unable to synthesize Golgi-modified complex N-linked glycans. *Plant Physiol.* 102: 1109–1118.

6 Scale-Up of Plant-Derived Pharmaceuticals
Prospects for Commercial Production and for Global Health

6.1 INTRODUCTION

In the past, biopharmaceuticals were produced chiefly by bacterial or yeast fermentation systems or mammalian cell culture. More recently, transgenic animals have been used for the production of biopharmaceuticals (Larrick and Thomas, 2003). The introduction of plants as an expression platform for commercialization of biopharmaceuticals is in its infancy. There are a number of very good reasons why plants make superior production platforms for biopharmaceuticals than conventional microbial fermentation; these points are illustrated in Table 6.1. First and foremost, as plants are higher organisms, they possess an endomembrane system. Plant cells are capable of folding and assembling recombinant proteins in a manner resembling those found in mammalian cells and as a result, they perform similar posttranslational modifications. In this way, antibodies and other multimeric complexes that are biologically functional have been expressed in various plant tissues (see Chapter 5).

In addition to this, health risks that stem from contamination with human pathogens become minimized when one is working with plants. There are no concerns of contamination with infectious agents such as prions or mammalian viruses. Ethical considerations that are involved with working with animals can be avoided. Another attribute of plant expression systems for commercial purposes is the fact that storage of recombinant proteins produced in plants is for the most part stable. Since transgenic plants can be recrossed or self-fertilized and can be maintained and propagated by traditional horticultural techniques, their seeds can also be stored and distributed (Giddings et al., 2000). The recombinant protein

TABLE 6.1A
Comparison of Expression Systems

Feature	Bacteria	Yeast	Mammalian Cell Culture	Transgenic Animal Products (Milk, Eggs)	Transgenic Plants (Field Crops)	Plant Cell Culture
Cost for production	Moderate	Moderate	Expensive	Expensive	Inexpensive	Inexpensive
Timescale speed of production	Fast	Fast	Fast	Slow	Fast	Fast
Safety	Issues of contamination	Issues of contamination	Issues of contamination infectious agents	Yes	Yes	Highly contained
Posttranslational modifications	No	Yes	Yes	Yes	Yes	Yes
Regulatory compliance	High	High	High	Complex	Complex	High
Yield	High	High	Moderate	Moderate	High	High
Scalability	Moderate	Moderate	Moderate	Slow	Rapid	Rapid
Protein stability	Yes	Yes	Yes	Yes	In seed, not fresh tissue	Yes

TABLE 6.1B

Comparison of Different Field Crop Species for Plant-Made Biopharmaceuticals

Crop	Advantage	Disadvantage
Tobacco	Easy to manipulate, large and rapid harvest	Toxic metabolites produced
Potato	Easily manipulated, can be stored for long periods	Requires cooking prior to consumption, destroys recombinant protein
Tomato	Grown rapidly, grown globally, can be contained in greenhouse, fruit can be eaten raw, can be freeze-dried	Fruit has a short shelf-life, more expensive to grow
Rice, maize	High expression of recombinant protein	Grows slowly
Banana	Inexpensive to grow, commonly grown in developing countries, fruit eaten raw, or as dried chips	Takes several years to produce a fruit, difficult to transform
Alfalfa, clover	Fast growing, large and rapid harvests, store seed for up to three years at room temperature	Alfalfa leaves contain oxalic acid, which might interfere with processing
Lettuce	Fast growing, eaten raw	Spoils quickly, protein becomes unstable
Soybean, pigeon pea	High biomass, very economical	Low expression levels, difficult to transform
Oilseed rape	Easy to extract protein using oleosin-fusion system	Low yields
Wheat, barley	Stability during storage	Low yields, difficult to transform
Arabidopsis	Easy to manipulate	Low biomass
Carrot, apple	Fast growing, eaten raw, in the form of carrot, apple juice, can be stored for long periods	

can be targeted to a particular plant organ or tissue (e.g., ER, chloroplast or vacuole, apoplastic space, seeds or tubers) where they can be stored for months or even years in some cases without substantial loss and with protein stability maintained.

Crops are also more amenable to upscaling or downscaling. For example, using the technologies currently available to produce biopharmaceuticals in plants, several crops can be grown on the same acreage per year, indicating that a virtually limitless amount of recombinant protein can be produced at a minimal cost. Plants are easy to grow and require neither special media or treatment nor the use of toxic chemicals as do fermentation systems (Fischer et al., 2004). Procedures to extract protein from plant tissues have already been developed, and are, in general, simple and inexpensive. Technology is already available for harvesting and processing plant material on a large scale. Since some of these plant products can be administered as food products such as edible vaccines, the time and expense of a purification process can be virtually eliminated. Since there is no need for fermentation equipment or skilled personnel to run it, production of recombinant proteins becomes as low as 2%–10% that of microbial fermentation systems and 0.1% that of the cost of mammalian cell culture systems.

The amount of vaccine or therapeutic protein produced in a plant can reach industrial-scale levels. The typical yield of biopharmaceuticals produced in a plant-based system is 0.1%–1.0% of total soluble protein (Twyman et al., 2003). This value is competitive with other expression systems; therefore, a plant expression platform for biopharmaceuticals is economically viable. As an example, one bushel of maize can produce as much avidin as one ton of chicken eggs but at 0.5% the cost.

Furthermore, any low protein yield can easily be compensated by the huge quantity of generated biomass. Fischer and coworkers calculated that 1 hectare of tobacco plants that produce an average sIGA yield of 0.5–1 g per kilogram fresh weight would result in 100 tons of fresh material per year, from which 50 kg sIgA could be harvested. Cost for protein is less than $50 per gram (Boehm, 2007). Analyses from other sources indicate that it would cost only $43/g to purify recombinant protein in corn but $105/g in transgenic goat's milk and $300–3000/g in mammalian cell cultures (Menkhaus et al., 2004).

A number of obstacles exist for marketing plant-made pharmaceuticals (PMP). For example, it is paramount that production of the PMP must be less expensive than for animal vaccines (e.g., vaccine presented in a form already in use as a feed). They must be able to successfully compete with a crowded marketplace dominated by microbial fermentation.

6.2 PLANT-MADE PHARMACEUTICALS ON THE MARKET

To date, a handful of protein products have been produced in plants; all have a diagnostic function rather than a direct function in human health. Products currently on the market include the following: Avidin, a chicken glycoprotein found in egg whites, can be used as a diagnostic reagent, and is produced in transgenic corn at a concentration of 20% total soluble protein. Avidin can be readily processed and purified, and is currently sold by Sigma-Aldrich (Hood et al., 1999; Kusnadi et al., 1998). β-Glucuronidase, also known as GUS, is another diagnostic reagent produced in transgenic corn and marketed by Sigma-Aldrich (Evangelista, 1998). Both of the above products cater to a relatively small market. The first large scale protein product from transgenic plant technology is trypsin (TrypZean™), also sold by Sigma-Aldrich. The production of maize-derived trypsin, a protease used for processing of biopharmaceuticals, is very attractive due to its significant market potential (the worldwide market for trypsin in 2004 was $120 million). Corn-derived trypsin is also attractive because it is free of the biological pathogens/contaminants that are often copurified and difficult to eliminate using other production platforms. The enzyme is expressed in an inactive zymogen form in corn, and the active enzyme becomes reconverted upon extraction from the cornmeal flour (Woodard et al., 2003).

The first commercial products for human use are expected to reach the market by 2009 (Sparrow et al., 2007). The first registration for a plant-derived vaccine was granted to Dow AgriSciences in January of 2006; this was for a poultry vaccine against Newcastle disease, and was produced in tobacco cell culture. In addition to this, Cuba's Biotechnology and Genetic Engineering Center won approval in April of 2006 to produce monoclonal antibodies against hepatitis B virus in whole tobacco plants.

6.3 SCALE-UP OF PRODUCTION OF BIOPHARMACEUTICALS IN PLANTS

6.3.1 PRODUCTION VIA OPEN FIELDS

There are two basic ways to express proteins in plants in an open field setting. Plants can be either stably transformed to express the desired protein, or alternatively, proteins can undergo transient expression via agroinfection or virus expression vectors (Rigano and Walmsley, 2005). It is easy to increase production of field grown crops by simply increasing the acreage (Table 6.1A). This unlimited scalability is of tremendous advantage, even under containment conditions, as it is possible to grow large amounts of crops under immense greenhouse facilities. For instance, 250 acres of greenhouse space would be sufficient to grow enough transgenic potato plants to satisfy the demand for hepatitis B virus vaccine in Southeast Asia. In many instances, a particular biopharmaceutical is required in large quantities, such as monoclonal antibodies used as topical mucosal treatments. For example, Guy's 13 MAb against *Streptococcus mutans* and HIV protein microbicides are very much needed at an inexpensive cost and in large quantities for the developing world (Rice et al., 2005). Both of these criteria can be met using field grown biopharmaceuticals as compared to other production platforms. Two examples of biopharmaceuticals produced on a large-scale basis to achieve levels of the protein suitable for commercialization are described below.

As a first example, bovine trypsin, as mentioned above, has a large market since it is widely used for digestion of other proteins (Woodard et al., 2003). Transgenic maize seedlings were grown in the greenhouse, flowered, and produced seed through hand pollination. After purification, 3.3% of total soluble protein or 58 mg trypsin/kg maize seed were produced. The trypsin produced from maize seed was determined to be functionally identical and physically similar to native bovine trypsin. At the same time, maize-derived trypsin was found to be free of pathogens and other pancreatic enzymes that often contaminate the purification process.

As a second example, Arlen et al. (2007) expressed type 1 interferon IFN-alpha2b in tobacco chloroplasts and transgenic lines, which were then grown in the field. IFN-alpha2b levels grown in this manner were determined to be biologically active and reached concentrations as great as 20% of total soluble protein, or 3 mg/g leaf (fresh weight). This represents the first field production of a plant-derived human blood protein.

6.3.2   Plant Species Enlisted for Open Field Production

To date, plant platforms under investigation for use in the large-scale production of biopharmaceuticals have focused on crop plants that are commonly targeted for human or animal use. Among these are tomatoes, potatoes, bananas, carrots, lettuce, maize, alfalfa, white clover, pigeon pea and *Arabidopsis*. A number of these plant species are listed below; others are described in Table 6.1B.

6.3.2.1   Plant Production Platforms for Human Health

Potato: Potato tubers have been used extensively as the crop of choice in over three separate human clinical trials so far and have been utilized to produce vaccines, human serum albumin, TNF-α and various antibodies. However, since transgenic potatoes expressing the desired biopharmaceutical must be delivered in an uncooked form, they have been known to cause gastrointestinal side effects in several volunteers used in clinical trials, and therefore will require some other forms of processing before they can be used widely.

Corn: Corn has been used extensively for the expression of biopharmaceuticals. A great advantage is the stability of the protein in the seed, which is in turn easy to store and transport.

Tomato: Tomato is also popular for the expression of biophamaceuticals due to the relative ease of processing of the fruit, and the fact that the fruit can be consumed fresh, thus providing stability of the recombinant protein at room temperatures. Tomato fruit has a short shelf life but, alternatively, can be freeze-dried, an inexpensive and well-established technology. Freeze-drying results in concentration of the protein and maintenance of batch consistency.

6.3.2.2 Plant Production Platforms for Animal Health

Alfalfa: Alfalfa has a high biomass yield and is a perennial plant that fixes its own nitrogen. Any glycoproteins that are synthesized in alfalfa leaves tend to have a highly homologous glycan structure. A disadvantage, however, is that alfalfa leaves contain oxalic acid, which may interfere with any downstream processing of the biopharmaceutical protein in question.

Safflower: In safflower plants expressing biopharmaceuticals, the protein of interest is fused to oleosin, the protein that forms oil bodies within the safflower seeds. The seeds can be crushed and the oil bodies then easily purified by centrifugation. This oleosin-fusion protein system was first developed by SemBioSys Genetics, Inc., in safflower or oilseed rape.

It should be noted that different species of field crops possess different biomass yields. For example, for every hectare of pea crop grown, only a third of a hectare of tomato crop would be required to produce the same amount of protein, even though the yield of protein per unit biomass is lower in tomato than in pea. Thus, consideration of which crop species to use becomes of paramount importance in the design of large-scale production of a particular biopharmaceutical. The relative biomass yields for a number of crops currently used for the production of biopharmaceuticals are listed in Table 6.2.

6.4 APPLICATION OF VIRUS EXPRESSION VECTORS FOR LARGE-SCALE FIELD TRIALS

A detailed description of plant virus vectors for expression of biopharmaceuticals is found in Chapter 4. The development of application of plant viral expression vectors for large-scale field trials involves many important issues. In order for production to be scaled up, it is necessary to have

TABLE 6.2

Biomass Yields of Various Crop Species

Plant	Portion Harvested	Tonnes/ha
Tobacco	Leaves	>100
Tomato	Fruit	68
Alfalfa	Leaves	25
Maize	Seed	12
Rice	Seed	>6
Wheat	Seed	3
Pea	Seed	2.4

Data derived from Daniell, H., Streatfield, S.J., and Wycoff, K. (2001). *Trends in Plant Science* 6(5): 219–226 and Schillberg, S., Twyman, R.M., Fischer, R. (2003). *Vaccine* 23: 1764–1769.

economically viable, pure and biologically active products that are free from both contaminants and infectious agents (Turpen et al., 1995; Pogue et al., 2002). In the case of plant-virus based systems, one key issue is the ability of the foreign insert sequence to be maintained through several passages in plants. Mutations in the foreign insert may render the product of less therapeutic or immunogenic value, and genetic drift becomes a problem. Several studies have been conducted to measure genetic drift associated with foreign sequences within plant viral expression vectors. These studies revealed that the viral sequences are in general quite stable, and the viruses can readily produce homogeneous populations of proteins in plants (Modelska et al., 1998; Wu et al., 2003). Deletion of all or large portions of the foreign sequence have been known to exist. Various pressures on viral replication and movement lead to the accumulation of deletion events in these nonviral sequences. In order to prevent loss of foreign gene sequences in the case of recombinant viral vectors that contain duplicate subgenomic promoters, heterologous promoters can serve as the subgenomic promoter and therefore improve stability of the vector (Wagner et al., 2004; Monger et al., 2006).

Other issues regarding the large-scale application of plant viral-based vectors include their ability to infect a wide spectrum of hosts, as well as to tolerate a variety of adverse environmental conditions and host defence responses. To assist in the replication of the virus at optimal levels under these conditions in the field, new varieties of host plants (transgenic and nontransgenic alike) that will allow vigorous virus infection must be developed; work is currently under way.

In the case of large-scale field applications, the initial production of large amounts of genetically stable primary virus expression vectors is required. Virus production takes place using a highly susceptible host, such as *N. benthamiana*. To ensure optimal growth of plants in the field, the appropriate agronomic procedures must also be practiced for each type of host plant (e.g., correct plant spacing, row structure, the use of fertilizers and watering schedule; Peres-Filgueira et al., 2003; Turpen et al., 1995).

In the case of recombinant viruses that no longer retain the ability to move systemically throughout plants, alternative methods of delivering the virus vector besides inoculation are required. Recently, the agricultural biotech company Icongenetics, Inc. has developed a novel way for transfecting plants with recombinant virus vectors, which they term "magnifection" (Toth et al., 2002). The best results were obtained by immersing an entire plant into a bacterial suspension and applying a weak vacuum (0.5–1 bar) for 1–2 min, followed by a gentle (<1 min) gradual return to atmospheric pressure. This results in the infiltration of *A. tumefaciens* suspension into the intercellular space of all mature leaves of tobacco or *N. benthamiana* plants. Plants that were so treated were then returned to the greenhouse. Heterologous protein expression was observed in all leaves. This novel procedure has many advantages as the vectors are less affected by transgene size since the simultaneous infection of multiple cells in multiple leaves means that infection is more synchronous and faster than systemic infection. As a result, the vectors do not have to move systemically, and a larger proportion of the host plant is infected by the virus (with systemic infection, only the young leaves are infected).

In 2007, Azhakanandam et al. developed an alternative amplicon plus targeting technology to produce protein at high levels in plants. In this case, they expressed L1 protein of canine oral papillomavirus by expressing a transgenic tobacco plant expressing a suppressor or gene silencing (enhances systemic infection of tobacco) with a PVX expression vector expressing L1 that is targeted to the chloroplasts (subcellular organelle that is protective for highly labile proteins). They used an *Agrobacterium*-inoculation "wound and spray" technique in which plants are wounded very slightly, then sprayed with an *Agrobacterium* suspension containing the plasmid of interest. The result is high yield in the short time (as little as 1–2 weeks) required for harvest. This technique is beneficial for producing large amounts of highly labile proteins at a very high throughput.

The lack of human pathogenicity of plant viruses rules out the risks of human infection by exposure in the field or in food products to a plant virus. However, biological containment of the virus expression vector remains a primary safety concern as it can be considered a risk to the environment. This includes the spread of recombinant viruses to weeds or nearby crops and the assessment of recombinant products on target and nontarget organisms. Different approaches can be employed for each plant expression vector. For example, TMV is spread by mechanical inoculation and not by insects or pollen. The standard practice to control TMV infection is by crop rotation and planting TMV-resistant plants surrounding the test plants. When the field trial is terminated, all plant tissue in the field is destroyed. Genes that confer virus spread by insects, fungi, nematodes, etc., can be removed in other instances. HC-Pro, for example, a gene product that is responsible for aphid transmissibility of potyviruses, can be removed or mutated from potyviral vectors. Safety containment strategies for plant expression vectors that have been transfected into the host plant via "magnifection" are based upon the fact that the replicon RNAs do not express CP and are therefore not packaged into virus particles, thus resolving biosafety issues for this system.

6.5   PRODUCTION OF BIOPHARMACEUTICALS IN PLANT SUSPENSION CELLS

Suspension cells are produced from larger aggregates of undifferentiated cells known as calli and are generated by treating leaf, stem, and root tissues with a variety of hormones that induce proliferation. Calli are cultivated on solidified media; the callus is then transferred to liquid medium and agitated on rotary shakers or in fermenters, resulting in single cell cultures. Plant cell cultures can be cultivated in normal shaker flasks, conventional fermenters with a few minor adjustments, or in large controlled bioreactors (Fischer et al., 1999). Plant cells have been cultivated up to a volume of 100,000 L. Cell suspension cultures are considered to be an excellent system for the production of biopharmaceuticals under good manufacturing practices (GMP), a very important asset for proteins that are being prepared for clinical use. The most popular cell lines, BY-2 (derived from Bright Yellow tobacco cultivar) and NT-1 (derived from *Nicotiana* tobacco 1) are preferred because of their ease of propagation and growth rate characteristics (Takeda et al., 1992). Both NT-1 and BY-2 cells have lost the ability to regenerate and constantly produce high levels of protein but have the added advantage of displaying a low alkaloid content (Santi et al., 2006). At present, a number of suspension cells derived from various crop species have been developed. A variety of antibodies and proteins have been produced in shaker flasks or fermentation cultures in tobacco cells, as well as pea, wheat and rice cell cultures (see Table 6.4). In all cases, these cell cultures can be cultivated on simple, inexpensive media and are capable of synthesizing complex multimeric proteins such as immunoglobulins with eukaryotic posttranslational modifications (Shin et al., 2003).

TABLE 6.3

Industries Involved in Commercialization of Biopharmaceuticals from Plants*

Name of Company	Location	Examples of Biopharmaceutical(s) Under Production	Examples of Plant Production Technology Platforms
Biolex, Inc.	Pittboro, North Carolina	Alpha-interferon, Anti CD20 antibodies, plasmin	Duckweed
Agracetus	Middleton, Wisconsin	Therapeutic proteins	Cotton, soy
Dow AgriSciences	Indianapolis, Indiana	Animal vaccines	Plant cell culture
Applied Phytologics	Sacramento, California	Alfa-1-antitrypsin	Rice
Croptech	Blacksburg, Virginia	Glucocerebrosidase	Undisclosed
Chlorogen, Inc.	St. Louis, Missouri	Human vaccines	Tobacco
Icon Genetics	Halle, Germany	Pharmaceuticals for human, animal health	Tobacco
Plantigen	London, Ontario	GAD, IL-10, IL-4, MHC, cytokines	Tobacco, potato
Epicyte Pharmaceuticals	San Diego, California	Alpha-herpes MAb	Maize
Greenovation, Inc	Freiburg, Germany	ADCC antibody	Moss cell line
Phytomedics	Jamesburg, New Jersey	Human alkaline phosphatase	Root system
Medicago, Inc.	Quebec City, Canada	H5N1 avian influenza vaccine	Alfalfa
Large Scale Biology Company	Vacaville, California	Various single chain Fv antibody fragments	Viral vectors in tobacco
LemnaGene	Lyon, France	Aprotinin	*Spirodela oligorrhiza* (type of duckweed)
Novoplant	Gatersleben, Germany	Antibody fragments against animal pathogens	Peas

—continued

TABLE 6.3 (continued)
Industries Involved in Commercialization of Biopharmaceuticals from Plants

Name of Company	Location	Examples of Biopharmaceutical(s) Under Production	Examples of Plant Production Technology Platforms
Planet Biotechnology	Hayward, California	Monoclonal antibody against dental caries	Tobacco
Meristem Therapeutics	France	Lactoferrin (for gastrointestinal infections), gastric lipase	Maize
Saponin, Inc.	Saskatoon, Saskatchewan	Saponins, anticancer agents, drugs against cardiovascular disease	Prairie carnation
PlantGenix	Malvern, Pennsylvania	Lycopene, lutein	Various plants
SunGene (BASF Plant Science)	Gatersleben, Germany	Plants with enhanced nutritional content	Canola, potato, *Arabidopsis*
SembioSys Genetics, Inc.	Calgary, Alberta	Insulin, ApoAI	Safflower seeds
Prodigene, Inc.	College Station, Texas	Vaccine against *E. coli* LT-B, trypsin, GUS, avidin, aprotinin, collagen, Brazzein (natural protein sweetener), TGEV vaccine in swine	Transgenic maize
Unicrop, Ltd.	Helsinki, Finland	Monoclonal antibodies, anticancer drugs	False flax
Ventria BioSciences	Fort Collins, Colorado	Lactoferrin, lysosome	Rice, barley
Cobento Biotech A/S	Denmark	Human intrinsic factor, for vitamin B12 deficiency	*Arabidopsis*

* Note: Information was Derived from Company Websites.

The calli used to generate cell lines can be transgenic and express the vaccine or therapeutic protein of interest. Transfer of foreign genes into calli takes place by *Agrobacterium*-mediated transformation, particle bombardment, electroporation of protoplasts, or by viral vectors (Fischer

TABLE 6.4
Examples of Bioreactors Expressing Plant-Made Biopharmaceuticals

Cell Culture	Protein Produced	Reference	Amount Produced	Notes
Rice suspension culture	Alpha(1)-antitrypsin rAAT	Trexler et al., 2005, Huang et al., 2001	3–12 mg/day, 200 mg/L	Cyclical, semicontinuous culture system
Rice	rAAT	McDonald et al., 2005	100–247 mg/L	Two compartment membrane bioreactor
Tobacco cell culture	Anti-rabies virus monoclonal antibodies	Girard et al., 2006	0.5 mg/L	Disposable plastic bioreactor
Tobacco NT-1 suspension cells	Human secreted alkaline phosphatase	Becerra-Arteaga et al., 2006	27 mg/L	Batch culture
Tobacco NT-1 suspension cells	IL-2, IL-4	Magnuson, 1998	8–180 µg/L	Shaker flasks
Tobacco suspension cells	IL-12	Kwon et al., 2003	800 µg/L	Shaker flasks
BY-2 cells	Thrombomodulin	Gao et al., 2004	27 µg/g	Shaker flasks
Rice suspension cells, tomato suspension cultures	Human granulocyte-macrophage colony stimulating factor (hGM-CSF)	Hong et al., 2006, Kwon et al., 2003	2.5 mg/L, 0.7% total secreted protein, 45 µg/L	Used rice alpha-amylase promoter (Ramy3D) system
Shaker flasks *Laminaria japonica* (kelp) gametophytes	Plasminogen activator protein	Gao et al., 2005	56 µg/g dry cell weight	Illuminated bubble column bubble reactor
Tobacco hairy roots	GFP	Modina-Bolivar et al., 2004	800 µg/L	Plastic sleeve bioreactor
NT-1 suspension cells	HBsAg	Smith et al., 2002	2 mg/L	Shaker flasks

et al., 1999). Recombinant proteins can then either be found in the culture supernatant or retained within the cells. If the protein is located within the cell itself, additional equipment and labor are required for its purification. The preferred approach is to target recombinant proteins to be secreted outside of the cell and into the liquid culture media. In this way, cell material can be removed by centrifugation followed by clarification of the filtrate prior to any purification step. If the protein of interest is produced within the plant cell, then a cell disruption step is required. This is often accomplished by sonication, and the bulk cell material is then removed by centrifugation and clarification steps (Fischer et al., 1999).

Plant cell culture has been used extensively for the production of biopharmaceuticals. A few examples are illustrated here, and a more extensive list can be found in Table 6.4. Smith et al. (2002) employed both soybean and tobacco cell lines to produce a hepatitis virus surface antigen (HBsAg), to be used as a vaccine, in shaker flask cultures. The authors found that the titers of HBsAg in soybean cell culture were 65 µg/g fresh weight, and 10-fold lower in tobacco cell culture, resulting in productivities of 1 mg/L/d and 0.16 mg/L/d, respectively. These numbers correspond well with those found for yeast batch cultures (1.5 mg/L/d).

Trexler et al. (2005) used rice suspension cultures to produce alpha (1)-trypsin (rAAT) using a cylindrical, semicontinuous strategy in which the rice cells were induced to secrete trypsin into the medium, which could then be repeatedly exchanged. Expression of rAAT is under control of a rice alpha amylase promoter that is induced in the absence of sugar. This cycle was repeated three times over a 25–28 day period. Using this technique, 3 to 12 mg of trypsin could be produced per day. Alternatively, McDonald et al. (2005) used a two-compartment membrane bioreactor for the production of rAAT in rice cells. The recombinant protein product is retained in the cell compartment, and later withdrawn from the bioreactor as part of a clarified solution. The authors were able to produce 4%–10% of total extracellular protein 5–6 days postinduction.

6.6   COMPARISON OF CELL CULTURE OVER WHOLE PLANTS FOR BIOPHARMACEUTICAL PRODUCTION

One of the greatest advantages to the use of cell culture over that of open field for production of biopharmaceutcials is that issues such as containment or regulatory approval no longer have to be addressed. Variations in soil and weather prohibit GMP conditions that are indispensible for pharmaceutical production for field grown plants. On the contrary, cell suspensions can be grown in precisely controlled environments suitable for GMP. In particular, when the protein can be secreted from cells, the cells secreting the product can be grown continuously, resulting in less expensive downstream processing. Recently, the first plant-derived pharmaceutical, a poultry vaccine for Newcastle disease that was produced in plant cell culture, received approval for market release (Dow Agrisciences, Indianapolis, IN; G. Walsh, 2006). On the other hand, plant cell cultures also exhibit certain disadvantages when compared to field grown plants. One disadvantage is the genetic instability of cell cultures over a period of subculturing. To circumvent this problem, Schmale et al. (2006) have developed a cryopreservation protocol for BY-2 suspension cells expressing human serum albumin. The authors were able to show that growth and productivity remain the same after cryopreservation. Cryo-cell banking will be important for any planned industrial usage of suspension cells for production of biopharmaceuticals.

Further comparisons of plant-made biopharmaceutical production in cell culture versus whole plants have been examined in a number of studies (Schinkel et al., 2005). In one approach, thrombomodulin was found to be expressed in whole tobacco plants at 0.1%–2.5% of total soluble protein (TSP), which was 4–5 times higher than in BY-2 suspension cultures (Gao et al., 2004). Another study exhibited different results, however. Solulin™ thrombomodulin was found to be produced in tobacco plants, at levels as great as 115 µg/g of fresh weight in transgenic tobacco leaves, or at 1.5% of TSP. In BY-2 cells, the maximum yield was determined to be 2.7 µg/g fresh weight, or 0.4 % of TSP. In this instance, the higher production of the recombinant protein was found in cell culture rather than in transgenic plants. Similarly, Girard et al. (2006) developed a cell culture from a transgenic tobacco plant that produced anti-rabies virus monoclonal antibodies at levels 3 times exceeding that of the original transgenic plant. Similar production rates were found when the cells were grown in Erlenmeyer flasks or in a disposable plastic bioreactor.

6.7 OTHER EXPRESSION SYSTEMS FOR LARGE-SCALE PRODUCTION OF BIOPHARMACEUTICALS

6.7.1 HAIRY ROOT CULTURE AND RHIZOSECRETION

The use of hairy roots for the production of biopharmaceuticals has been studied extensively and has been discussed in Chapter 1 of this book. To date, over 116 different plant species have been induced to produce hairy roots in culture (Guillon, 2006). Originally, an expression system was developed for protein production based on the natural secretion from roots of intact plants. In order to take up nutrients from the soil, interact with other soil organisms, and defend themselves against numerous pathogens, plant roots have developed sophisticated mechanisms based upon the secretion of different biochemicals in the rhizosphere. Compounds, including recombinant proteins, can then move through the roots via the intercellular space (apoplast) and have direct contact with the external environment. Using this concept, studies were conducted using various marker proteins to show that root secretion can be exploited for the continuous protein production in a process called *rhizosecretion*. By fusing proteins to signal peptides, which localizes them to the endoplasmic reticulum, proteins can be targeted to the root apoplast through the plant's own secretion pathway by rhizosecretion (Borisjuk et al., 1999).

Rhizosecretion is easy to scale up and very cost effective with respect to isolation and purification. However, the bioreactor systems used for hairy root cultures differ from those used for plant cell suspensions. Traditional bioreactor systems have recently been adapted for root culture, and this technology is now being taken to commercial scales. The most traditional system is the airlift bioreactor used for microorganisms or plant cells. This system is adapted for the culturing of roots in liquid medium. Mist culture systems have also been developed. For this technology, the volume of the culture medium is reduced and the concentration of the secreted therapeutic protein is increased. If the protein to be produced is known to be quite stable, then a less expensive hydroponic culture can be designed in a manner suitable for scale-up.

The production of pharmaceutical proteins using hairy roots and rhizosecretion technology represents a safe and viable alternative to the use of whole plants for molecular pharming. As an example of the efficiency of this system, Medina-Bolivar and Cramer (2004) expressed the reporter protein GFP in tobacco hairy root cultures using a plastic sleeve bioreactor with a 5 L volume. Yields of 500 µg GFP/L after 21 days of incubation or 20% of total secreted protein were produced using this expression system, suggesting that rhizosecretion offers a promising production system for the production of biopharmaceuticals.

6.7.2   MOSS

Moss (*Physcornitrella patens*) has also been examined for its potential as a production host for biopharmaceuticals (Cove, 2005). Moss possesses many of the advantages of plant cell suspension systems: precise control over growth conditions, batch-to-batch product consistency, high levels of containment, and ability to comply with good manufacturing practices. Moss can be grown on relatively simple media consisting of inorganic salts, with airborne CO_2 as the sole carbon source. Moss, like plant suspension cells, can be grown inexpensively in agitated glass flasks, photo-bioreactors, stirred glass tanks, or in modular, fully scalable reactors. However, moss does not display the same degree of genetic instability as is found in some plant cell cultures.

The moss genome has been fully sequenced, rendering it an easy species to genetically engineer. The codon usage pattern in moss allows expression of human genes without any requirement for codon adaptation. In addition, the pathways for posttranslational modification such as N-glycosylation are the same as in higher plants. Moss "knockout" and "knockin" strains have now been developed that possess the correct human glycosylation expression patterns. Moreover, the relative strengths of heterologous promoters from both plants and animals have been quantified in moss in addition to a number of constitutive and inducible promoters that are specifically derived from moss. In a similar fashion, a number of secretion signals have now been evaluated in moss (Decker and Reski, 2007).

PEG-mediated transfection of moss protoplasts is the transformation procedure routinely used for this production host. Harvest of the recombinant protein can take place within hours to several weeks. Recently, Decker and Reski (2008) took advantage of the ease of manipulation of the moss genome and were able to coexpress a protection agent in a transgenic moss line, which further enhanced recovery of the target protein during the extraction procedure. This particular strain of moss was able to yield a high recovery of vascular endothelial growth factor (VEGF), which could then be harvested and concentrated to 10 μg/mL via continuous product separation. In fact, in another study, moss-derived monoclonal antibodies were shown to be superior to mammalian Chinese hamster cell-derived antibodies. Moss bioreactors, therefore, offer a safe and efficient scaleable system for the production of complex biopharmaceuticals, and the commercial feasibility of this production platform is under consideration by a number of pharmaceutical companies (Hellwig et al., 2004).

6.7.3 Aquatic Plants

Other plant expression systems that have potential for production of biopharmaceuticals as another alternative to open field plants are a number of free-floating aquatic plant species. These include both higher plants, such as duckweed and kelp, as well as several species of algae. In all cases the systems are very cost effective due to their fast growing abilities. As a result, bulk production can take place either in contained in vitro cultures, within greenhouses or in ponds. For example, Gao et al. (2004) generated the transgenic kelp *Laminaria japonica*, expressing plasminogen activator protein. The kelp was maintained in a culture in an illuminated bubble column bioreactor (see Table 6.4).

For algae expression systems, *Chlamydomonas reinhardtii* and *Chlorella ellipsoidea* have been under investigation. However, synthetic genes must be designed in this case because algae possess a codon usage pattern that differs from other plants. To date, chloroplast-targeted transgenes have expressed an antibody against glycoprotein D of HSV in the green algae *Chlamydomonas reinhardtii* (Boehm, 2007). Higher plants that have been examined as suitable candidates for molecular pharming include duckweed (Lemnaceae) and *Spirodela oligorrhiza*. Duckweed has an additional advantage as it can also be used as animal feed, thus limiting the requirement for downstream processing as found with other edible crops such as tomato (Boehm, 2007).

6.8   DOWNSTREAM PROCESSING OF PLANT-DERIVED BIOPHARMACEUTICALS

The extraction and purification of proteins from organisms or biological tissue can be a laborious and expensive process, and often represents the principal reason why vaccines and other therapeutic agents reach costs that become unattainable for many. Downstream processing also can be a major obstacle with respect to cost for large-scale protein manufacturing in plants. However, purification from plant tissues, while still costly, is in general less expensive than purification from their mammalian and bacterial counterparts. Indeed, some plant-derived biopharmaceuticals, such as topically applied monoclonal antibodies, may require only partial purification and thus be even less intensive in terms of labor and cost.

A series of common steps are required from the harvest of the crop to the acquisition of purified protein (see Figure 6.1). In general, crop harvest is followed by storage and at least partial processing of the plant material (as in the case of corn kernels/ground cornmeal or wheat grain/flour). Processing may include extraction of soluble plant material or oil, depending on the tissue type that the recombinant protein is predicted to accumulate in. The next series of steps involve protein extraction and clarification from plant tissue. Removal of waste plant material by centrifugation is often included at this stage. The final stage involves more intricate purification procedures, such as immunoprecipitation and chromatography. Some plant types contain harmful alkaloids and care must be taken to keep them out of the purification process (S[-] nicotine in tobacco, for example).

Purification of a recombinant therapeutic protein from cell culture may involve a few simple steps. For example, affinity chromatography may be used to concentrate the protein. This can be followed with an ion-exchange step and gel filtration.

A number of purification strategies are in place that have been tailored for specific proteins derived from particular host production platforms. For example, downstream protein extraction from seeds is quite simple

Harvest Crop

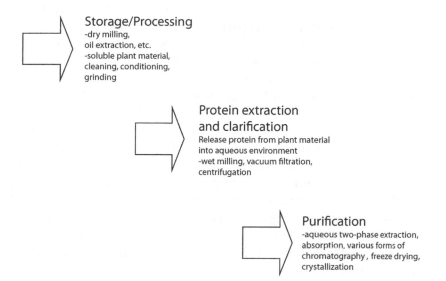

Storage/Processing
-dry milling,
oil extraction, etc.
-soluble plant material,
cleaning, conditioning,
grinding

**Protein extraction
and clarification**
Release protein from plant material
into aqueous environment
-wet milling, vacuum filtration,
centrifugation

Purification
-aqueous two-phase extraction,
absorption, various forms of
chromatography, freeze drying,
crystallization

FIGURE 6.1   Common steps required from the harvest of the crop to the acquisition of purified protein.

when compared to extraction from vegetable or leaf tissue. One of the most promising new technologies is that of hirudin, an anticoagulant used to treat thrombosis, and utilized for commercial production by SemBioSys, Calgary, Canada. Using this strategy, the *hirudin* gene is fused to the *oleosin* gene in transgenic oilseed rape. An endoprotease cleavage site between the two proteins is included in the construct, which then becomes activated upon recovery of the hirudin. Oleosin-hirudin fusion proteins are isolated from oil bodies, which can be separated from plant tissue by flotation centrifugation (Kühnel et al., 2003). Using this particular method of purification, the hirudin becomes active only after being harvested from the plant tissue.

Another technique to ease purification involves the utilization of a peptide sequence derived from elastin-like protein (ELP). By adding 100 repeats of ELP to their protein of interest, in this case a single-chain antibody which accumulated in tobacco seed, Lin et al. (2006), were able to use a technique known as *inversion transition cycling*. This technology takes advantage of the fact that ELP-fusion proteins become soluble in water below their transition temperature, and become insoluble when the temperature is raised. The authors generated transgenic tobacco plants expressing sgp130, the membrane glycoprotein used as a therapeutic agent in combating Crohn's disease, rheumatoid arthritis, and colon cancer, as a sgp130-ELP fusion protein. Inverse transition cycling led to large amounts of sgp130 accumulation. This protein could then be purified easily and was demonstrated to be functionally active. Recombinant proteins have also been expressed in plants as part of a fusion construct containing an integral membrane spanning domain from the human T-cell receptor. This domain enables the recombinant protein to be purified from the membrane fraction of plant tissue (Schillberg et al., 2005).

Extraction techniques have been developed for the purification of recombinant proteins from maize. For example, aprotinin, a protease inhibitor produced in transgenic maize, has been extracted from cornmeal using a trypsin-agarose column, with a recovery of 49% (Azzoni et al., 2002). Well-established commercial methods have now been developed for the separation of maize germ from endosperm; if a protein of interest can be targeted to one of these locations, purification becomes greatly simplified. As a result of this knowledge, in a later study dry milling was used to separate the germ and endosperm fractions of transgenic maize expressing aprotinin (Zhong et al., 2007). The germ fraction was identified as containing the aprotinin, and further chromatography steps were conducted on this fraction, resulting in a much higher recovery of 75.3%.

Similarly, trypsin has been purified from ground maize seed flour using ST1-agarose, followed by cation exchange chromatography.

Monoclonal antibodies are also routinely and easily purified from plants using a routine procedure. For example, Pujol et al. (2005) constructed an scFv antibody fragment against the hepatitis B virus surface antigen (HBsAg) with a histidine-tag in tobacco plants. The scFv was demonstrated to be biologically active and could be readily purified to 95% using immobilized metal-ion affinity chromatography. The yield was determined to be 15–20 µg/g fresh leaves produced, a value comparable to the yield in *E. coli* of 20 µg/L. Tobacco cell suspensions infected with transformed *A. tumefaciens* processed in a similar fashion resulted in a yield of only 0.3–0.5 µg/mL.

His-tagged GUS-fusion proteins have been produced and isolated from tobacco chloroplasts. His-tagged proteins have also been extracted by foam fractionation (Crofcheck et al., 2003, 2004) or by a modified intein expression system (Morassutti et al., 2002).

In another processing procedure, Platis and Labrou (2006) purified anti-HIV MAb from plants using a PEG/Pi aqueous two-phase system. This technique purifies the plant extract from any additional alkaloids and phenolics. The sample can then be applied to a protein A or G affinity column without the need for any further treatment. The authors were able to use this technique to achieve over 95% recovery.

In a more recent example, a technique has been developed to purify anti-HIV monoclonal antibodies from tobacco leaf extracts and maize seed (Platis and Labrou, 2008). The authors found histamine selectively bound to the MAb, allowing a high degree of purification and recovery in a single step. The authors also found a plant-host dependency of their purification procedure: MAb from tobacco bound and purified at a reduced capacity compared to maize extract.

In addition, monoclonal antibodies have been produced using the ELP fusion and inversion transition cycling technique described before by Floss et al. (2008). MAb directed against HIV-1 and coupled to ELP exhibited enhanced stability and simplified recovery, while at the same time displaying no deleterious effects on biological activity, assembly, and folding of the monoclonal antibody.

6.9 DRIVING FORCES BEHIND PLANT-MADE BIOPHARMACEUTICALS: THE GENERAL MARKET ENVIRONMENT

6.9.1 COMMERCIALIZATION OF PLANT-DERIVED PRODUCTS AND THE CORPORATE ARENA

The considerable list in Table 6.3 illustrates some of the companies that are currently preparing plant-based biopharmaceutical technologies for commercialization. Steps have been taken toward the development and subsequent commercialization of one of these products, currently in Phase 2 clinical trials and known as CaroRx™ (licensed to Planet Biotechnology Inc., Hayward, CA); a monoclonal antibody designed for treatment of tooth decay by *S. mutans* is described in the following text.

The development of Guy's 13 secretory IgA plantibody technology commenced with the work of Ma et al. (2005), and involved the sexual crossing of four transgenic plants, each expressing heavy and light immunoglobulin domains, as well as the J chain and secretory component. Plants were screened that could express all four proteins simultaneously. It was determined that assembly of all four proteins into a single macromolecule was highly efficient in transgenic tobacco plants. Furthermore, functional studies showed that this plant produced sIgA bound specifically to its native antigen, and in a clinical trial was able to effectively prevent oral colonization by *S. mutans* via passive immunization of the mucosal surfaces. This represented the first demonstration of a therapeutic protein produced in plants that had a clinical application in humans.

Similar to the work being done at Planet Biotechnology, Inc., a number of small biotech companies are aiming toward commercializing antibodies produced from plants (Berghman et al., 2005). It has been estimated that the expected annual expansion based on the requirement of sIgA is 13%

and estimated revenue will increase by $25 billion per year (Berghman et al., 2005). Furthermore, the estimated cost of monoclonal antibodies from crops is significantly lower than for other production systems. For example, a crop such as maize can produce sIgA at approximately $0.1/g. The cost would be $300/g in animal cell culture, $1–2/g from transgenic animals (milk, eggs) and $1/g from bacterial fermentation. In addition to this, since antibodies can be expressed in several crops that are commonly a part of human or animal diets, no downstream purification would be necessary in many instances. This alone represents 95% of the production cost. Of course, many other benefits exist, as mentioned previously, such as ease of storage and administration, protection of the antibody from degradation, and its gradual release into the gut through bioencapsidation within edible plant tissue.

While this all sounds like a dream come true, significant problems still exist with respect to public perception and the concept of molecular farming, which force pharmaceutical companies developing plans to use plants as production platforms to take careful note. The general lack of enthusiasm among the pharmaceutical industry community stems from issues regarding working with GM-related foods and the possible social repercussions that these issues may generate. Other considerations, such as new regulatory guidelines that may need to be developed regarding the particular plant species, manner of containment, efficiency, and cost of downstream processing procedure, etc., lower the confidence of investors for return on their investment. As a result, it has been considered by many that plant-derived biopharmaceuticals will initially have more success penetrating the animal health market, rather than the human health market.

6.9.2 PHILANTHROPIC USES FOR PLANT-DERIVED PHARMACEUTICALS

One original driving force for molecular farming has been to develop new vaccines and therapeutic agents that target the most devastating infectious diseases found in developing countries. Diarrhea, the major cause of mortality globally, and other diseases that prevail in developing countries, are not being prioritized by the private sector, as there is little hope of return on investment. The ease of administration, low cost, and requirement for partial processing has made the concept of plant-derived vaccines, antibodies, and other therapeutic agents feasible for providing relief to third world countries. The fact remains that 20% of the world's infants are still left unimmunized, and infectious diseases are responsible for 2 million preventable deaths a year, due to constraints on vaccine production, distribution, and delivery. For example, ETEC, the enterotoxin released during infection by some strains of *E. coli*, is responsible for 3 million infant deaths a year, predominantly in impoverished areas. Administration of a plant-derived vaccine against ETEC to mothers might be useful in immunizing the fetus in utero by transplacental transfer of maternal antibodies or to the infant by breast milk. Plant-derived vaccines would be useful against those diseases that are rare and whose cures are not well financed, such as dengue fever, hookworm, and rabies. Plant-based vaccines can also be incorporated into other vaccine programs, using vaccines in a number of different formats in conjunction with conventional vaccines and as primary immunizations or boosters, as is described in the next chapter on mucosal immunity. In addition to this, multicomponent vaccines can be made in plants to target several infectious diseases that are major killers in the third world, such as cholera, rotavirus, yellow fever, polio, malaria, and HIV. Adjuvants can also be coexpressed along with the vaccine protein in the same plant combination to assist in eliciting a strong mucosal immune response.

In order to develop the technology and ensure that plant-derived vaccines have an impact on those in the third world who do not have access to such medicines, both governments and philanthropic organizations such as the Gate's Foundation and Ford Foundation must work together to overcome issues including the lack of investment for research and development for novel technologies to help the poor. Achieving global immunization, for instance, depends greatly on the cooperation of both local governments and international health organizations.

In 1992, the World Health Organization in Geneva and a consortium of other philanthropic organizations began the Children's Vaccine Initiative. The first attempt of this initiative was to lead to the production of transgenic potato tubers expressing hepatitis B surface antigen (discussed in detail in Chapter 2) to produce a vaccine against hepatitis B virus for developing countries. Ultimately, the target host plant to be used is the banana, which grows throughout the developing world and, unlike potatoes, can be eaten raw. Banana vaccines are offered as a puree to maintain consistency of the vaccine protein; these would cost only a small percentage of the price of its conventional vaccine counterpart. Using this type of strategy, a mere 40-acre plot has been projected to produce enough hepatitis B vaccine in plants to vaccinate all of China each year, and a 200-acre plot could vaccinate all babies of the world (Giddings, 2000).

Another example of putting plants to work for philanthropic purposes is the case of rabies virus, a disease that causes approximately 55,0000 deaths a year in Southeast Asia and Africa but does not receive significant financial attention because it is not as significant a killer as some other diseases. The vaccine currently available is largely too expensive for developing countries. Transgenic plants could produce antibodies against rabies virus inexpensively and in large quantities (Ma et al., 2005). The vaccine could be grown locally, would not require manufacturing facilities, could be stored at the site of use and is heat stable (useful for the tropics). Syringes or needles would not be required for administration, lowering the chance of infection or contamination from infectious agents such as mad cow disease (Lai et al., 2007). A large-scale production system would be highly economical due to the low technology requirements involved. Since the cost of downstream processing of protein can make up a substantial amount of the total production cost, this can be minimized by choosing a host species that is amenable to its location of use, and of which processing is only partially required. The relatively affordable infrastructure and set-up costs may be amenable to developing countries themselves (Schillberg et al., 2005). The vaccine protein could also be propagated through greenhouses, or cell culture bioreactors. Plant tissue can also be freeze dried or ground and further processed, and then administered in the form of a tablet or capsule for oral delivery.

A number of problems remain regarding these plans to immunize children in developing countries with vaccine and other therapeutic proteins produced in edible plant tissue. For example, the exact dose of a particular vaccine administered to a child is critical. Too low a dose may fail to induce immunity, while too high a dose may cause tolerance. It is only realistic to predict that some children may spit out some of their banana containing vaccine protein, thus reducing the dosage and rendering the vaccination ineffective. It may therefore be necessary to consider providing the vaccine protein in the form of a baby food, or banana chips, puddings, or pills to assure consistency and that the child receives the correct dosage.

Another significant obstacle to be overcome is the impending bottlenecks that are encountered in managing intellectual property for cooperative research and development programs of a philanthropic nature. To be granted the freedom to operate for various entities, including foundations, to develop technologies can be a tall order, depending on the circumstances. In the same vein, the organization of intellectual property to enable products to be developed in partnership with private companies and foundations can also be complex and difficult. Efforts need to be taken to mitigate such hurdles.

6.9.3 PLANT-DERIVED BIOPHARMACEUTICALS AS A DEFENSE AGAINST BIOLOGICAL WARFARE AGENTS

An increased awareness of the potentials of global bioterrorism has also prompted the investigation of nonconventional ways to create inexpensive vaccines on a rapid, massive scale if necessary (Hilleman, 2002). Plant-derived vaccines offer a potential new defense against biological warfare agents as they are inexpensive and can be easily scaled up to millions of doses within a short time frame. The vaccine can be stored for several years at ambient temperatures in the form of corn grain, or administered to troops or to the general public in small tomato ketchup packets, as examples. Efforts have been made to generate plant production systems against a number of potential biological warfare agents, such as *Staphylococcus* enterotoxin B, PA of anthrax and plague antigen (Azhar et al., 2002; Hefferon and Fan, 2004; Alvarez et al., 2005). These are discussed in more detail in Chapters 2, 4, and 7, respectively.

6.10 CONCLUSIONS

At present, the production capacity of the pharmaceutical industry is overwhelmed. This inability of the corporate sector to keep up with demands may result in more attention being drawn toward the use of plants for the production of biopharmaceuticals. While great progress has been made with respect to the large-scale production of plant-made biopharmaceuticals, much basic research is still required to pave the way for commercialization of these products. Problems remain such as difficulties with low protein yield, the possible deleterious effects on protein activity/function due to differences in glycosylation patterns, and the potential impact of plants expressing biopharmaceutical proteins on the environment (for example, concerns surrounding gene containment). Indeed, many proteins are expressed in plants at such low concentrations that it is not yet economical to produce them at an industrial scale. Thus, a variety of practical considerations including biomass yield, the recovery of recombinant protein per unit biomass, the ease of transformation and in vitro manipulation, product storage conditions, downstream purification and processing strategies, market size, environmental concerns, public perception, competing technologies and the ability to scale up the product to make the biopharmaceutical protein available to the target population are all factors that must be evaluated in great detail. For example, the choice of plant expression system currently used depends very much on the type of protein and its applications. As technologies improve for increasing yields, production can be extended to an even broader range of plant species.

REFERENCES

Alvarez, M.L., Pinyerd, H.L., Crisantes, J.D., Rigano, M.M., Pinkhasov, J., Walmsley, A.M., Mason, H.S., and Cardineau, G.A. (2005). Plant-made subunit vaccine against pneumonic and bubonic plague is orally immunogenic in mice. *Vaccine* 24(14): 2477–2490.

Arlen, P.A., Falconer, R., Cherukumilli, S., Cole, A., Cole, A.M., Oishi, K.K., and Daniell, H. (2007). Field production and functional evaluation of chloroplast-derived interferon-alpha2b. *Plant Biotechnol. J.* 5(4): 511-525.

Azhakanandam, K., Weissinger, S.M., Nicholson, J.S., Qu, R., and Weissinger, A.K. (2007). Amplicon-plus targeting technology (APTT) for rapid production of a highly unstable vaccine protein in tobacco plants. *Plant Mol. Biol.* 63: 393–404.

Azhar, A.M., Singh, S., Anand Kumar, P., and Bhutnagar, R. (2002). Expression of protective antigen in transgenic plants: a step towards an edible vaccine against anthrax. *Biochem. Biophys. Res. Commun.* 299(3): 345–351.

Azzoni, A.R., Kusnadi, A.B., Miranda, E.A., and Nikolov, Z.L. (2002). Recombinant aprotinin produced in transgenic corn seed: extraction and purification studies. *Biotechnol. Bioeng.* 80(3): 268–276.

Berghman, L.R., Abi-Ghanem, D., Waghela, S.D. and Ricke, S.C. (2005). Antibodies: an alternative for antibiotics? Symposium: antibiotics in animal feed: are there viable alternatives? 660–666.

Boehm, R. (2007). Bioproduction of therapeutic proteins in the 21st century and the role of plants and plant cells as production platforms. *Ann. N.Y. Acad. Sci.* 1102: 121–134.

Borisjuk, N.V., Borisjuk, L.G., Logendra, S., Petersen, F., Gleba, Y., and Raskin, I. (1999). Production of recombinant proteins in plant root exudates. *Nat. Biotechnol.* 17(9): 466–469.

Cove, D. (2005). The Moss *Physcomitrella patens. Annu. Rev. Genet.* 39, 339–358.

Crofcheck, C., Loiselle, M., Weekley, J., Maiti, I., Pattanaik, S., Bummer, P.M., and Jay, M. (2003). Histidine tagged protein recovery from tobacco extract by foam fractionation. *Biotechnol. Prog.* 19(2): 680–682.

Crofcheck, C., Maiti, I., Pattanaik, S., and Jay, M. (2004). Effect of ion and surfactant choice on the recovery of a histidine-tagged protein from tobacco extract using foam fractionation. *Appl. Biochem. Biotechnol.* 119(1): 79–92.

Daniell, H., Streatfield, S.J., and Wycoff, K. (2001). Medical molecular farming: production of antibodies, biopharmaceuticals and edible vaccines in plants. *Trends Plant Sci.* 6(5): 219–226.

Decker, E.L. and Reski, R. (2007). Moss bioreactors producing improved biopharmaceuticals. *Curr. Opin. Biotechnol.* 18(5): 393–398.

Decker, E.L. and Reski, R. (2008). Current achievements in the production of complex biopharmaceuticals with moss bioreactors. *Bioprocess. Biosyst. Eng.* 31(1): 3–9.

Evangelista, R.L., Kusnadi, A.R., Howard, J.A., and Nikolov, Z.L. (1998). Process and economic evaluation of the extraction and purification of recombinant beta-glucuronidase from transgenic corn. *Biotechnol. Prog.* 14(4): 607–614.

Fischer, R. Emans, N., Schuster, F., Hellwig, S., and Drossard, J. (1999). Towards molecular farming in the future: using plant-cell-suspension cultures as bioreactors. *Biotechnol. Appl. Biochem.* 30(Pt. 2): 109–112.

Fischer, R., Stoger, E., Schillberg, S., Christou, P., and Twyman, R.M. (2004). Plant-based production of biopharmaceuticals. *Curr. Opin. Plant Biol.* 7: 152–158.

Floss, D.M., Sack, M., Stadlmann, J., Rademacher, T., Scheller, J., Stöger, E., Fischer, R., and Conrad, U. (2008). Biochemical and functional characterization of anti-HIV antibody-ELP fusion proteins from transgenic plants. *Plant Biotechnol. J.* Epub ahead of print.

Gao, J., Hooker. B.S., and Anderson, D.B. (2004). Expression of functional human coagulation factor XIII A-domain in plant cell suspension s and whole plants. *Protein Exp. Purif.* 37(1): 89–96.

Giddings, G., Allison, G., Brooks, D., and Carter, A. (2000). Transgenic plants as factories for biopharmaceuticals. *Nat. Biotechnol.* 18: 1151–1156.

Girard, L.S., Fabis, M.J., Bastin, M., Courtois, D., Petiard, V., and Koprowski, H. (2006). Expression of a human anti-rabies virus monoclonal antibody in tobacco cell culture. *Biochem. Biophys. Res. Commun.* 345(2): 602–607.

Guillon, S.M., Tremoulliaux-Guiller, J., Kumar Pati, P., Rideau, M., and Gantet, P. (2006). Harnessing the potential of hairy roots: dawn of a new era. *Trends Biotechnol.* 24(9): 403–409.

Hefferon, K.L. and Fan, Y. (2004). Expression of a vaccine protein in plants using a geminivirus-based replicon system. *Vaccine* 23(3): 404–410.

Hellwig, S., Drossard, J., Twyman, R.M., and Fischer, R. (2004). Plant cell cultures for the production of recombinant proteins. *Nat. Biotechnol.* 22(11): 1415–1422.

Hilleman, M.R. (2002). Overview: cause and prevention in biowarfare and bioterrorism. *Vaccine* 20(25–26): 3055–3067.

Hood, E.E. and Jilka, J.M. (1999). Plant-based production of xenogenic proteins. *Curr. Opin. Biotechnol.* 10(4): 382–386.

Hood, E.E., Kusnadi, A., Nikolov, Z., and Howard, J.A. (1999). Molecular farming of industrial proteins from transgenic maize. *Adv. Exp. Med. Biol.* 464: 127-147.

Kuhner, B., Alcantara, J. Boothe, J., van Rooijen, G., and Moloney, M. (2003). Precise and efficient cleavage of recombinant fusion proteins using mammalian aspartic proteases. *Protein Eng.* 16(10): 777-783.

Kusnadi, A.R., Hood, E.E., Witcher, D. R., Howard, J.A., and Nikolov, Z.L. (1998). Production and purificationof two recombinant proteins from transgenic corn. *Biotechnol. Prog.* 14(1): 149-155.

Kwon, T.H., Kim, Y.S., Lee, J.H., and Yang, M.S. (2003). Production and secretion of biologically active human granulocyte-macrophage colony stimulating factor in transgenic tomato suspension cultures. *Biotechnol. Lett.* 25(18): 1571–1574.

Lai, P., Ramachandran, V.G., Goyal, R., and Sharma, R. (2007). Edible vaccines: current status and future. *Indian J. Med. Microbiol.* 25(2): 93–102.

Larrick, J.W. and Thomas, D.W. (2003). Producing proteins in transgenic plants and animals. *Curr. Opin. Biotechnol.* 12: 411–418.

Lin, M., Rose-John, S., Götzinger, J., Conrad, U., and Scheller, J. (2006). Functional expression of a biologically active fragment of soluble gp130 as an ELP-fusion protein in transgenic plants: purification via inverse transition cycling. *Biochem. J.* 398: 577–583.

Ma, J.K.-C, Barros, E., Bock, R., Christou, P., Dale, P.J., Dix, P.J., Fischer, R., Irwin, J., Mahoney, R., Pezzotti, M., Schillberg, S., Sparrow, P., Stoger, E., and Twyman, R.M. (2005). Molecular farming for new drugs and vaccines. *EMBO Rep.* 6(7): 593–599.

Magnuson, N.S., Linzmaier, P.M., Reeves, R., An, G., HayGlass, K., and Lee, J.M. (1998). Secretion of biologically active human interleukin-2 and interleukin-4 from genetically modified tobacco cells in suspension culture. *Protein Exp. Purif.* 13(1): 45–52.

Medina-Bolivar, F. and Cramer, C. (2004). Production of recombinant proteins by hairy roots cultured in plastic sleeve bioreactors. *Methods Mol. Biol.* 267: 351-363.

Menkhaus, T.J., Bai, Y., Zhang, C., Nikolov, Z.L., and Glatz, C.E. (2004). Considerations for the recovery of recombinant proteins from plants. *Biotechnol Prog.* 20(4): 1001–1014.

Modelska, A., Dietzschold, B., Fleysh, N., Fu, Z.F., Steplewski, K., and Hooper, C. (1998). Immunization against rabies with plant-derived antigen. *Proc. Natl. Acad. Sci. U.S.A.*, 95: 2481–2485.

Monger, W., Alamillo, J.M., Sola, I., Perrin, Y., Bestagno, M., Burrone, O.R., Sabella, P., Plana-Duran, J., Enjuanes, L., Garcia, J.A., and Lomonossoff, G.P. (2006). An antibody derivative expressed from viral vectors passively immunizes pigs against transmissible gastroenteritis virus infection when supplied orally in crude plant extracts. *Plant Biotechnol J.* 4(6): 623–631.

Perez-Filgueira, D.M., Zamorano, P.I., Dominguez, M.G., Taboga, O., Del Medico Zajac, M.P., Puntel, M., Romera, S.A., Morris, T.J., Borca, M.V., and Sadir, A.M. (2003). Bovine herpes virus gD protein produced in plants using a recombinant tobacco mosaic virus (TMV) vector possesses authentic antigenicity. *Vaccine* 21: 4201–4209.

Platis, D. and Lebrou, N.E. (2006). Development of an aqueous two-phase partitioning system for fractionating therapeutic proteins from tobacco extract. *J. Chromatogr. A* 1128(1–2): 114–124.

Platis, D., and Labrou, N.E. (2008). Affinity chromatography for the purification of therapeutic proteins from transgenic maize using immobilized histamine. *J. Sep. Sci.* 31(4): 636–643.

Pogue, G.P., Lindbo, J.A., Garger, S.J., and Fitzmaurice, W.P. (2002). Making an ally from an enemy: plant virology and the new agriculture. *Annu. Rev. Phytopathol.* 40: 45–74.

Pujol, M., Ramírez, N.I., Ayala, M., Gavilondo, J.V., Valdés, R., Rodríguez, M., Brito, J., Padilla, S., Gómez, L., Reyes, B., Peral, R., Pérez, M., Marcelo, J.L., Milá, L., Sánchez, R.F., Páez, R., Cremata, J.A., Enríquez, G., Mendoza, O., Ortega, M., and Borroto, C. (2005). An integral approach towards a practical application for a plant-made monoclonal antibody in vaccine production. *Vaccine* 23(15): 1833–1837.

Rice, J., Ainsley, W.M., and Showen, P. (2005). Plant-made vaccines: biotechnology and immunology in animal health. *Anim. Health Res. Rev.* 6(2): 199–209.

Rigano, M.M. and Walmsley, A.M. (2005). Expression systems and developments of plant-made vaccines. *Immunol. Cell Biol.* 83: 271–277.

Santi, L., Huang, Z., and Mason, H. (2006). Virus-like particles production in green plants. *Methods* 40(1): 66–76.

Schillberg, S., Twyman, R.M., and Fischer, R. (2005). Opportunities for recombinant antigen and antibody expression in transgenic plants-technology assessment. *Vaccine* 23: 1764–1769.

Schinkel, H., Schiermeyer, A., Soeur, R., Fischer, R., and Schillberg, S. (2005). Production of an active recombinant thrombomodulin derivative in transgenic tobacco plants and suspension cells. *Transgenic Res.* 14(3): 251–259.

Schmale, K., Rademacher, T., Fischer, R., and Hellwig, S. (2006). Towards industrial usefulness—cryo-cell-banking of transgenic BY-2 cell cultures. *J. Biotechnol.* 104(1): 302-311.

Shin, Y.J., Hong, S.Y., Kwon, T.H., Jang, Y.S., and Yang, H.S. (2003). High level of expression of recombinant human granulocyte-macrophage colony stimulating factor in transgenic rice cell suspension culture. *Biotechnol. Bioeng.* B82(7): 778–783.

Smith, M.L., Mason, H.S. and Shuler, M.L. (2002). Hepatitis B surface antigen (HbsAg) expression in plant cell cultures: kinetics of antigen accumulation in batch culture and its intracellular form. *Biotechnol. Bioeng.* 80(7): 612–619.

Sparrow, P.A., Irwin, J.A., Dale, P.I., Twyman, R.M., and Ma, J.K. (2007). Pharma-Planta: road testing the developing regulatory guidelines for plant-made pharmaceuticals. *Transgenic Res.* 16(2): 147–161.

Takeda, Y., Hirokawa, H., and Nagata, T. (1992). The replication origin of proplastid DNA in cultured cells of tobacco. *Mol. Gen. Genet.* 232(2): 191-198.

Thomas, B.R., van Deynze, A., and Bradford, K.J. (2002). *Production of Therapeutic Proteins in Plants.* Agricultural Biotechnology in California Series, Publication 8078.

Trexler, M.M., McDonald, K.A., and Jackman, A.P. (2005). A cyclical semicontinuous process for production of human alpha 1-antitrypsin using metabolically induced plant cell suspension cultures. *Biotechnol. Prog.* 21(2): 321-328.

Turpen, T.H., Reinl, S.J., Charoenvit, Y., Hoffman, S.L., Fallarme, V., and Grill, L.K. (1995). Malarial epitopes expressed on the surface of recombinant tobacco mosaic virus. *Biotechnology*, 13: 53–57.

Twyman, R.M., Stoger, E., Schillberg, S., Christou, P., and Fischer, R. (2003). Molecular farming in plants: host systems and expression technology. *Trends Biotechnol.* 21(12): 570–579.

Wagner, B., Fuchs, H., Adhami, F., Ma, Y., Scheiner, O. and Breiteneder, H. (2004). Plant virus expression systems for transient production of recombinant allergens in *Nicotiana benthamiana. Methods* 32: 227–234.

Walsh, G. (2006). Biopharmaceutical benchmarks. *Nat. Biotechnol.* 24: 769–776.

Woodard, S.L., Mayor, J.M., Bailey, M.R., Barker, D.K., Love, R.T., Lane, J.R., Delaney, D.E., McComas-Wagner, J.M., Mallubhotla, H.D., Hood, E.E., Dangott, L.J., Tichy, S.E., and Howard, J.A. (2003). Maize (*Zea mays*)-derived bovine trypsin: characterization of the first large-scale, commercial protein product from transgenic plants. *Biotechnol. Appl. Biochem.* 28(Pt. 2): 123–130.

Wu, X., Dinneny, J.R., Crawford, K.M., Rhee, Y., Citovsky, V., Zambryski, P.C., and Weigel, D. (2003). Modes of intercellular transcription factor movement in the Arabidopsis apex. *Development* 130: 3735–3745.

Zhong, Q., Xu, L., Zhang, C., and Glatz, C.E. (2007). Purification of recombinant aprotinin from transgenic corn germ fraction using ion exchange and hydrophobic interaction chromatography. *Appl. Microbiol. Biotechnol.* 76(3): 607–613.

7 The Immune Response to Plant-Derived Pharmaceuticals

7.1 INTRODUCTION

Much attention in the field of plant-made pharmaceuticals has so far focused on the generation of vaccines and other therapeutic proteins that can be produced inexpensively and applied for various medical applications. Many of these plant-derived vaccines have been designed to prioritize diseases that enter via the mucous membrane and are major causes of mortality in developing countries. Heat-stable plant-made vaccines that are administered orally, therefore, have the potential to enhance vaccine coverage in children and infants, particularly in resource-poor regions of the world. In particular, plant-based vaccines delivered orally are well suited for combating gastrointestinal diseases; this has been the focus of several phase 1 clinical trails in humans and animals (Thananala et al., 2006).

The induction of an immune response by various mucosal routes is an important approach for the control of mucosally acquired infections. The apparent linked nature of the mucosal immune system enables the delivery of an antigen to any mucosal surface to have the secondary effect of potentially inducing immunity at others. Induction of a combination of systemic and secretory immune responses can be determined by the nature of the antigen, the route of administration, and the delivery system utilized. For example, traditional parenteral vaccines primarily induce IgM and IgG responses, whereas mucosal vaccination can elicit both IgG and secretory IgA responses (Corthesy, 2007).

The fundamental advantage to be gained by the use of plant-derived vaccines is the ability to deliver protein immunogens of various pathogens to the sites where the immune response is active in the gut (Streatfield, 2005a, 2005b). One major obstacle for the delivery of proteins to the intestinal immune system is the fact that many antigens cannot survive the harsh environment of the digestive tract. An advantage of plant-made vaccines that mitigates this hurdle is that plant tissues provide protection and prevent antigen degradation, while at the same time acting as a delivery vehicle for the antigen as it passes through the gut (Streatfield, 2005b). Another problem frequently encountered is that many antigens do not become recognized by the gut as foreign and therefore do not serve as immunogens. In this instance, adjuvants may be required to alter the immunogenic context in which an antigen is encountered. It is for this reason that many plant-based vaccine proteins are directly linked to a strong adjuvant that, for example, is produced simultaneously in the plant.

This chapter first provides a description of immunity in general and then more specifically, immunity in the mucosal immune system. The immune response of both intestinal and respiratory tracts will be described in detail as these are the two most common portals of targeted vaccine development for mucosal immunity. The chapter will cover the basis of mucosal immunity using plant-based oral vaccines. Strategies for increasing mucosal immunity, such as the use of adjuvants, will also be discussed. Finally, the chapter will cover the preclinical tests and various clinical trials that are taking place with respect to production of human and veterinary therapeutic proteins in plants.

7.2   THE IMMUNE SYSTEM IN GENERAL

The immune system can be defined as a combination of sophisticated mechanisms that protect an organism against infection. This process involves the sequential identification and distinguishing of pathogens (viruses, bacteria, parasites, or worms) from normal cells, followed by the killing of the invading pathogens. One pathway, known as the innate immune response, provides an immediate, nonspecific response to a pathogen upon exposure. The innate immune system is found in both plants and animals. In animals it consists of white blood cell recruitment, cytokine production, and other mechanisms to protect the host in a nonspecific fashion. Cell-mediated immunity involves the activation of macrophages, natural killer cells that destroy intracellular pathogens and cytotoxic T-lymphocytes that can undergo apoptosis (programmed cell death) and stimulate cells to release various cytokines (a protein used in cellular communication). Humoral immunity, on the other hand, involves the production and secretion of antibodies and the accessory processes that accompany it, such as cytokine production and memory cell generation (Fazilleau et al., 2007).

In jawed vertebrates, the immune system consists of many tissue types that interact together in an elaborate and dynamic network. In these organisms, the immune system has the capacity to adapt over time to recognize and create a specific memory against pathogens, and as a result offer more extensive protection, known as acquired immunity. The concept of vaccination is based upon this acquired immunity. The adaptive immune response requires the recognition of specific antigens during a process called *antigen presentation*. This antigen specificity allows for a response that is tailored to specific pathogens, which are then maintained by memory cells. Prominent in the adaptive immune system are lymphocytes, known as B and T cells, which carry receptor molecules that recognize specific targets. The next section discusses the role of T cells, B cells, and antibodies in the immune response. The reader may find useful Table 7.1, which provides definitions for much of the terminology used in this section.

7.3 T CELLS

T cells are named after the thymus, the paramount organ for T cell development. T cells can be divided into a number of subsets; these include cytotoxic killer T cells and helper T cells (Figure 7.1A). These cells recognize antigens that have been processed and presented along with a major histocompatibility complex (MHC) molecule. Cytotoxic T cells are activated when the T cell receptor binds to a specific processed antigen in a complex with the MHC (class I) receptor of another cell such as a B cell or dendritic cell. This complex recognition is assisted by the CD8+ coreceptor on the T cell. Activated killer T cells release cytotoxins and initiate apoptosis of the target cell (Lu, 2006: Fazilleau et al., 2007: Twigg, 2005).

Helper T cells, on the other hand, recognize antigen bound to class II MHC molecules, along with the CD4+ coreceptor located on the surface of helper cells. Helper T cells release cytokines that in turn regulate the activity of many cell types, including macrophages and killer T cells (Figure 7.1a). Helper T cells can differentiate into two subsets of cells. Th1 cells, or type 1 helper T cells, produce the cytokines γ-IFN (gamma-interferon) and β-TNF (beta-tumor necrosis factor). Th1 cells also stimulate the cellular immune response by stimulating macrophages and increasing CD8+ T cell proliferation. Th2 cells (type 2 helper T cells) produce the cytokines interleukin-4 (IL-4), IL-5, IL-6, IL-10, and IL-13. Th2 cells stimulate the humoral immune system by stimulating B cell proliferation and increasing antibody production (Twigg, 2005; Reiner, 2007; Woodfolk, 2007; Botturi et al., 2007).

7.4 B CELLS

B cells are named after the *bursa of Fabricus*, the organ located in birds from which B cells mature. B cells mature with the bone marrow in mammals. B cells can be separated into two main classes: plasma cells, which constitutively secrete antibodies, and memory cells, which secrete antibodies only in response to reexposure to particular antigen (Figure 7.1b). B cells can be further classified into B1 and B2 cells. B1 cells were first

TABLE 7.1

A Who's Who of the Immune System

Name	Other Names, Abbreviations	Role in Immune System
Macrophage	mø	Derived from monocytes, these highly phagocytic cells engulf and digest pathogens
Natural killer T cell	NK T cell	Subset of T cells, upon activation rapidly release large quantities of cytokines
Cytokine	None	Cellular-signaling molecules, glycoproteins that are secreted by various immune cells
Chemokine	None	Small proteins, subset of cytokines
Major histocompatibility complex	MHC	Classes of proteins that display pieces of foreign antigen on cell surface
T-helper cell	T_h	Subset of lymphocytes that activates or directs other immune cells
Cytotoxic T cell	T_C	Subset of lymphocytes that kills infected cells
CD4+	Cluster of differentiation 4	Transmembrane glycoprotein expressed on surface of some immune cells, interacts with antigen-presenting cell
CD8+	Cluster of differentiation 8	Transmembrane glycoprotein expressed on surface of cytotoxic T cell, required for activation
Interferon	IFN	Type of cytokine, activates many types of immune cells
Tumor necrosis factor	TNF	Cytokine that regulates immune cells
Interleukin	IL	Group of cytokines produced by various immune cells; promotes differentiation of T and B cells
Lymphocyte	None	Type of white blood cell, includes T cells, B cells, and NK cells
Plasma cell	None	Type of B cell; secretes large amounts of antibodies
Dendritic cell	DC	Found in close contact with the external environment, function as antigen-presenting cell
Mucosa-associated lymphoid tissue	MALT	System of lymphoid tissue found throughout the body such as the skin, GI tract, and lungs

TABLE 7.1 (continued)
A Who's Who of the Immune System

Name	Other Names, Abbreviations	Role in Immune System
Gut-associated lymphoid tissue	GALT	Immune system of the digestive tract
Peyer's patches	None	Aggregations of lymphoid tissue found in the lowest portion of the small intestine
Minifold cell	M cell	Found in the epithelium of the Peyer's patches; delivers antigen to antigen-presenting cells
Polymeric immunoglobulin receptor	pIgR	Mediates transcellular transport of polymeric immunoglobulin molecules such as IgA
Lamina propria	None	A portion of the mucous membrane, a layer of tissue found under the epithelium; contains lymphoid tissue

identified in the gastrointestinal tract and do not require help by T cells for maturation. B1 cells undergo class switching to become IgA secreting cells in the *lamina propia*, the thin layer of tissue that lies underneath the epithelial layer of the mucosa. B2 cells, on the other hand, require T-cell help, this process is mediated mainly through the secretion of a number of interleukins (Twigg, 2005; Vadjy, 2006).

7.5 ANTIBODIES AND THE IMMUNE RESPONSE

In 1897, Paul Ehrlicin showed that antibodies were responsible for humoral immunity. Antibodies, also known as immunoglobulins, are Y-shaped glycoproteins. Five classes of antibodies, known as isotypes, exist in mammals: IgA, IgD, IgE, IgG, and IgM; each differs with respect to their biological properties (Figure 7.1c). Antibodies bind to a specific antigen and can function either by forming complexes with other antibodies and agglutination of antibody complexes, by priming macrophages for phagocytosis, by blocking viral receptors or by stimulating various aspects of the immune response (Twigg, 2005; Janeway et al., 1999; Roitt et al., 2002; Prlic and Bevan, 2006)

Antibody isotypes can change upon B-cell activation. Upon maturity, both IgM and IgD become expressed as membrane-bound forms on the B cell surface. Activation results in the production of antibody in secreted forms. Isotype switching takes place with some daughter cells to pro-

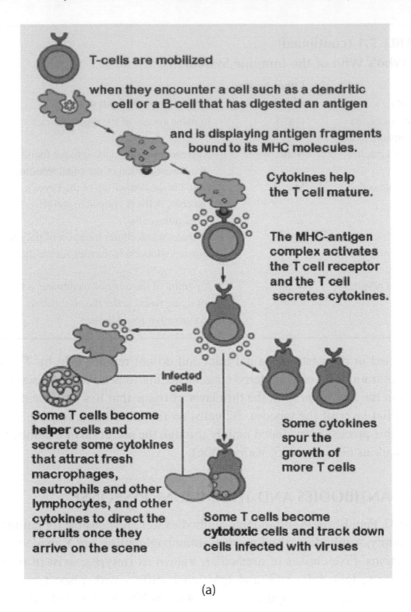

T-cells are mobilized

when they encounter a cell such as a dendritic
cell or a B-cell that has digested an antigen

and is displaying antigen fragments
bound to its MHC molecules.

Cytokines help
the T cell mature.

The MHC-antigen
complex activates
the T cell receptor
and the T cell
secretes cytokines.

Infected
cells

Some T cells become
helper cells and
secrete some cytokines
that attract fresh
macrophages,
neutrophils and other
lymphocytes, and other
cytokines to direct the
recruits once they
arrive on the scene

Some cytokines
spur the
growth of
more T cells

Some T cells become
cytotoxic cells and track down
cells infected with viruses

(a)

FIGURE 7.1(a) *A color version of this figure follows page 110.* Development of
cytotoxic and helper T cells. T cell maturation takes place upon activation of the
T cell by an antigen-presenting cell such as a dendritic cell or a B cell. Antigen
fragments are bound to the MHC complex and displayed on the cell surface of
an antigen-presenting cell. This MHC-antigen complex interacts with the T cell
receptor of the T cell and stimulates it to release cytokines. Cytokines secreted by
the activated T cell induce the proliferation of both cytotoxic T cells, which kill
infected cells, as well as helper T cells, which direct other immune cells at the
scene of infection.

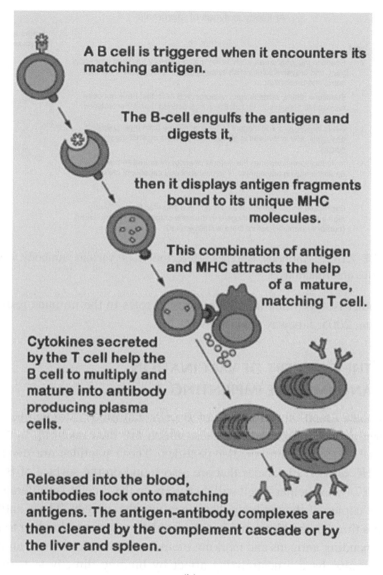

A B cell is triggered when it encounters its matching antigen.

The B-cell engulfs the antigen and digests it,

then it displays antigen fragments bound to its unique MHC molecules.

This combination of antigen and MHC attracts the help of a mature, matching T cell.

Cytokines secreted by the T cell help the B cell to multiply and mature into antibody producing plasma cells.

Released into the blood, antibodies lock onto matching antigens. The antigen-antibody complexes are then cleared by the complement cascade or by the liver and spleen.

(b)

FIGURE 7.1(b) *A color version of this figure follows page 110.* B cell activation and the generation of antibodies. Upon encountering an antigen, a B cell ingests the antigen and displays fragments of it upon its surface via a MHC complex. A T cell that interacts with the antigen-MHC complex of this B cell then releases cytokines to direct the B cell to proliferate and mature into antibody-producing plasma cells.

Antibody isotypes of mammals

Name	Types	Description	Antibody Complexes
IgA	2	Found in mucosal areas, such as the gut, respiratory tract and urogenital tract, and prevents colonization by pathogens. Also found in saliva, tears, and breast milk.	_
IgD	1	Functions mainly as an antigen receptor on B cells that have not been exposed to antigens. Its function is less defined than other isotypes.	Monomer IgD. IgE. IgG
IgE	1	Binds to allergens and triggers histamine release from mast cells and basophils, and is involved in allergy. Also protects against parasitic worms.	Dimer IgA
IgG	4	In its four forms, provides the majority of antibody-based immunity against invading pathogens. The only antibody capable of crossing the placenta to give passive immunity to fetus.	Pentamer IgM
IgM	1	Expressed on the surface of B cells and in a secreted form with very high avidity. Eliminates pathogens in the early stages of B cell mediated (humoral) immunity before there is sufficient IgG.	

FIGURE 7.1(c) Antibody isotypes. Description of various antibody isotype structures and functions.

duce other isotypes that have more defined roles in the immune response (Parham, 2005; Janeway et al., 1999)

7.6 THE CONCEPT OF VACCINATION AND IMMUNE IMPRINTING

Antibodies found on the surface of B cells can bind to foreign antigens to form antigen/antibody complexes, which are then taken up by the B cell and processed into smaller peptides. These peptides are displayed via MHC class II molecules that are also found on the surface of B cells (Figure 7.1B). A helper T cell will then activate the B cell to proliferate and secrete copies of the antibody that recognize the antigen. These antibodies pass through the blood and lymph circulatory systems, binding to their corresponding antigens and marking them for elimination. These antibodies may also have a neutralizing affect by binding directly to toxins or viruses and bacterial receptors (Ogra et al., 2001; Fazilleau et al., 2007).

A number of progeny B cells will become memory cells, which, upon repeated encounters with a specific pathogen, will then have the ability to mount a strong response on future challenges with that pathogen (Twigg, 2005; Vadjy, 2006). During vaccination (immunization), an antigen is taken from a particular pathogen to stimulate the immune system and to develop immunity without any of the pathogenic effects associated with that organism. Often, vaccines are based upon live attenuated viruses or components of bacteria that do not cause toxic effects but can induce an immune response. In many cases, adjuvants must be provided to maximize immunogenicity.

7.7 ORGANIZATION OF THE MUCOSAL IMMUNE SYSTEM

Mucosal surfaces are the site of infection and the portal of entry for most pathogens. The mucosal surfaces of the gut, airways, and urogenital tracts, as well as the ducts of the exocrine glands, are lined by epithelial layers that form tight barriers separating the rapidly changing external environment from highly regulated internal compartments. The mucosal surface is further protected by the innate and adaptive immune mechanisms described earlier, which bring about the recognition and eradication of pathogens (Clavel and Haller, 2007; Kang and Kudsk, 2007; Macperson and Uhr, 2004). Common to all mucosal sites is an epithelial surface that overlies organized lymphoid follicles. The mucosal epithelium contains mucin-producing glandular cells, lymphocytes, plasma cells, dendritic cells, and macrophages, as well as an assortment of cytokines and chemokines (Figure 7.2). This mucosa-associated lymphoid tissue (MALT) comprises a variety of sites required for antigen uptake, processing, and presentation for induction of mucosal responses.

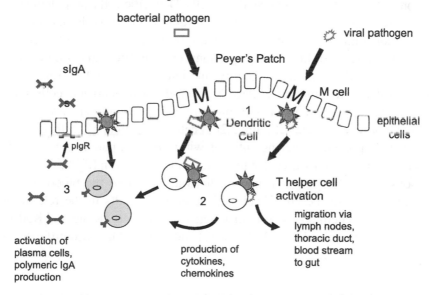

FIGURE 7.2 Immune response in the GALT. (1) Pathogens are delivered from the intestinal lumen to a type of antigen-presenting cell known as dendritic cells by M (minifold) cells. (2) Dendritic cells are then triggered to mature and migrate to lymphoid tissues where they present processed antigen to T cells and B cells. The degree of activation and migratory properties vary depending on the antigen and the environment in which the antigen is encountered, resulting in the production of different patterns of cytokines and chemokines. (3) T cells can then move via the thoracic duct into the bloodstream. Activated plasma cells produce polymeric IgA, which binds to pIgR and is exported across the epithelium as secretory IgA (sIgA).

The gut-associated lymphoid tissue (GALT) is the mucosal immune system specific to the intestinal tract. In vertebrates, the GALT represents 70% of the body's entire immune system. Peyer's patches (named after the 17th-century Swiss anatomist Hans Conrad Peyer) are the major sites of induction in mammals. Peyer's patches form large clusters of lymphoid follicles that are distributed along the length of the small intestine. In humans, the greatest density of Peyer's patches is found to aggregate within the ileum, or lowest portion of the small intestine. These appear as elongated thickenings of the intestinal epithelium and measure a few centimeters in length. Peyer's patches are involved in immune surveillance of the intestinal lumen and facilitate the mucosal immune response. Peyer's patches contain M (minifold) cells, highly specialized cells that deliver antigens from the lumen to antigen-presenting cells. M cells are important in luminal uptake, transport, and processing of mucosally introduced antigens. The differentiation and uptake processes of M cells remain poorly defined (Corthesy, 2007). Following exposure to antigen and its uptake via M cells is the activation of T cells, B cells, and dendritic cells. Dendritic cells are potent antigen-presenting cells that sample antigens from the intestinal lumen and are critical for initiating a primary immune response. Dendritic cells receive their name after their branched projections, or "dendrites" and are involved in controlling the maturation, quality, and intensity of the T cell response both locally and temporally. Peyer's patch-derived dendritic cells have unique features essential for both processing foreign antigens and subsequent immune imprinting (Figure 7.2). B cells and memory T cells become stimulated upon encountering antigens within Peyer's patches; these cells migrate and eventually arrive at the lymph nodes and amplify the immune response. Lymphocytes activated in this way pass into the blood stream and then travel to the gut. T cell activation also results in the release of cytokines and chemokines from different T cell subsets (Kang and Kudsk, 2007; Macpearson and Uhr, 2004; Clavel and Haller, 2007; Montufar-Sulis et al., 2007).

7.8   SECRETORY IgA AND THE MUCOSAL EPITHELIUM

The major antibody isotype in external secretions is sIgA, and the total amount of IgA synthesized is twice the amount of IgG produced daily in humans. IgA cells represent up to 80% of the entire mucosal lymphoid cell population. sIgA in mucosal secretions results from polymeric IgA transported across mucosal epithelium via binding to the pIgRreceptor (also known as the secretory component). The receptor is eventually cleaved and results in an IgA:pIgR complex, referred to as sIgA (Figure 7.2) (Rojas and Apodaca, 2002).

Activated T and B cells emigrate from the inductive environment via lymphatic drainage, circulate through the bloodstream, and target the mucosal effector sites. B cells expressing IgA differentiate during their spread to mucosal sites to become IgA-producing plasma cells. IgG also reaches these areas by passive diffusion from the bloodstream (Corthesy, 2007; Rojas and Apodaca, 2002).

IgA antibodies can play a number of roles in mucosal immunity. Luminal sIgA antibodies prevent adhesion and entry of antigens into the epithelium. IgA antibodies present in the *lamina propria* bind to and excrete antigen into the lumen. IgA antibodies transported through the epithelium can inhibit virus production or neutralize proinflammatory antigens. IgA also triggers the release of inflammatory mediators through receptors specific for its Fc domain. IgA antibodies can be generated in specialized lymphoid tissue just beneath the mucosal surface. Antigen-bound sIgA has also been demonstrated to specifically adhere to and be transported across M cells, interact with dendritic cells, and trigger an immune response (Corthesy, 2007).

7.9   MUCOSAL IMMUNE RESPONSE OF THE RESPIRATORY TRACT

In the respiratory tract, antigen is believed to be taken up in alveolar spaces by macrophages or other antigen-presenting cells. These cells migrate by lymphatosis to regional lymph nodes where the primary immune response takes place. Antigen-specific B cells are generated and travel back to the lung where they terminally differentiate and expand into either antibody secreting plasma cells or memory cells.

Studies indicate that the strongest mucosal immune response is induced when an antigen derived from a respiratory pathogen is introduced directly into the respiratory tract. Upon exposure to a potential pathogen, an antibody response in the respiratory tract can occur either quickly through the activation of resident memory B cells if there has been previous exposure to the pathogen, or more slowly through the induction of both systemic and local mucosal immunity if the host is naive to the pathogen. Antigen-specific IgG and IgA act in concert to help clear invading pathogens; the type and concentration of antibody produced is dependent on the site of exposure. Exposure of the upper airway results in primarily an IgA response. Organisms that pass through the airway and ultimately reach the lung induce a more systemic response, including an increased production of pathogen-specific IgG antibodies. Systemic vaccination against respiratory pathogens (which stimulates systemic IgG and elicits a modest mucosal IgA response) are less effective than vaccination through mucosal surfaces (which induce a brisk local and systemic IgA and IgG response), outlining the great importance regarding the development of vaccines that can be delivered as an aerosol by inhalation (Lu and Hickey, 2007; Ogra et al., 2001; Walker, 1994; Foss and Murlaugh, 2000).

7.10   ORAL TOLERANCE

Most substances that enter the gut are not immunogenic due to the cellular environment at the site of antigen presentation, as determined by cytokines as well as by other signals. This lack of response prevents unnecessary and damaging inflammatory responses to benign substances, which could lead to conditions such as inflammatory bowel syndrome or food allergies. Oral tolerance refers to the phenomenon of oral feeding with a specific protein and results in the abolishment of subsequent responses to systemic challenge with the same protein. This tolerance is a reflection of how the antigen is processed and presented to T lymphocytes in the mucosa. T cells play a role in mediating tolerance and a direct relationship has been shown between various cytokines and the development of mucosal tolerance (Faria and Weiner, 2006; Hajishengallis et al., 2005; Vajdy, 2006; Woodfolk, 2007).

7.10.1   ADJUVANTS AND OTHER STRATEGIES TO ENHANCE MUCOSAL IMMUNITY

One of the primary difficulties in developing a vaccine strategy is to induce an adequate mucosal immune response. It is often difficult to induce a mucosal response after administering an antigen because of the frequent inactivation of the antigen by both mucosal enzymes as well as bacterial flora (Ogra et al., 2001). As a result, the contact of antigen with mucosal tissues involved in antigen uptake or processing becomes a challenge. A number of approaches have been taken to stimulate the mucosal immune response and are listed in Table 7.2. The next section will deal with the role of mucosal adjuvants in stimulating the immune response to an antigen.

7.10.2   ADJUVANTS AND THE MUCOSAL SURFACE

Adjuvants are factors that increase immunogenicity of vaccine antigens. The term *adjuvant* come from the Latin *adjuvare*, meaning to assist or to help. The classic adjuvant was described by Freund in 1937 and consisted of paraffin oil and tubercle bacilli administered as a water-in-oil emulsion

TABLE 7.2

Strategies Employed to Increase Mucosal Immunity

Approach	Description	Examples
Recombinant vaccines	Vaccines that can deliver and penetrate intestinal wall via M cells	Tetanus toxin, Salmonella strains, vaccinia virus vector
DNA vaccines	Direct injection of nucleic acid contents	HIV, malaria, influenza, hepatitis B virus, cancer
Subunit vaccines	Immunogenic proteins or peptide antigens customized to specific antigenic determinants are purified from tissue culture	HIV, rabies virus, influenza virus, hepatitis B virus
Microcarrier particles	Can encapsulate antigen and be carried across the mucosal epithelium	Biodegradable microspheres, liposomes, virus-like particles, stimulating complexes (ISCOM[a])
Adhesive antigens	Highly efficient at inducing mucosal responses	LT toxin
Mucosal adjuvants	Bind to M cells and GM1 ganglioside receptors on mucosal epithelium	CT toxin

[a] Cage-like structures composed of glycosides.

along with the antigen. Later, Johnson et al. (1956) reported that purified lipopolysaccharides from various Gram-negative bacteria were capable of enhancing antibody formation.

One property of adjuvants is that they can increase intestinal immunogenicity by binding to epithelial cells (Chalmers, 2006). Adjuvanticity is mediated by signals from the innate immune system. The decision to respond or to not respond to a given antigen is made in a nonspecific manner and results from signals given from cells of the innate immune system via interleukins, cytokines, and costimulatory molecules to create potent antigen-presenting cells and initiate acquired immunity. For optimum effect, an adjuvant must display persistence by not being destroyed within the gut and by protecting antigens from degradation. It must be present in the correct location to enable specific binding to take place between epithelial cells to target antigens for uptake into the GALT. Adjuvants are also required to be in the correct environmental context to stimulate antigen presentation (Foss and Murtaugh, 2000; Lauterslager and Hilgers, 2002; Walker, 1994).

7.10.3   CT-B AS ORAL ADJUVANT

The prototype oral adjuvant is cholera toxin (CT), the secreted toxin of *Vibrio cholerae*. Cholera toxin is a member of the adenosine 5'-diphosphate (ADP)-ribosylating toxins and is derived from enterotoxigenic bacteria. These toxins induce a robust immunity when orally administered. Their structural features provide immunogenic and adjuvant stimulation in the intestinal mucosa. CT functions via the ADP-ribosylation of host G-proteins and as a result can modify the cellular environment for highly efficient antigen presentation. Heat-labile toxin (LT) of *Escherichia coli* is also an adjuvant. CT and LT possess 80% homology at the nucleotide level. LT, however, is not secreted by the bacteria; it is instead released by dead bacteria, and since it requires an additional protease cleavage event to take place to be active, it therefore is less effective as an adjuvant. CT is a multimeric protein that consists of five 11.6 kDa B subunits (CT-B) arranged in a pentameric ring and a single 27 kDa A subunit, which is located in the center of the ring (Figure 7.3A). CT-B subunits can bind to monosialoganglioside (GM_1) located on the surface of epithelial cells. CT-A, on the other hand, enters the cytosol and catalyzes the ADP-ribosylation of G-protein, which in turn elevates cyclic AMP and induces diarrhea (Walker, 1994; Ogra et al., 2001).

In humans, oral CT-B is immunogenic, resulting in both systemic and local antibody responses. CT can also act as an adjuvant for coadministered antigens. It should be noted that CT-B can act as an efficient transmucosal carrier molecule and delivery system for antigens (Daniell et al., 2001). Recently, CT-B was shown to deliver green fluorescent protein (GFP) as part of a GFP:CT-B fusion to the intestinal mucous membrane, via oral delivery to mice fed transgenic leaves expressing the fusion protein in chloroplasts. At this point, it is thought that GFP becomes cleaved from CT-B and crosses the intestinal lumen, suggesting that CT-B can be used successfully as a transmucosal carrier for the delivery of vaccines or other therapeutic proteins (Limaye et al., 2006). In this way, coupling proteins that are weakly immunogenic to CT-B can increase their antigenicity within the gut. CT-B can therefore increase uptake of the desired antigen and alter its delivery to the desired location. This, in turn, alters the context in which the antigen is encountered, directly with respect to antigen-presenting cells, and indirectly to a variety of cell types involved in the immune response. CT-B promotes presentation of the antigen of interest to antigen-presenting T cells in a context that aids in the stimulation of the

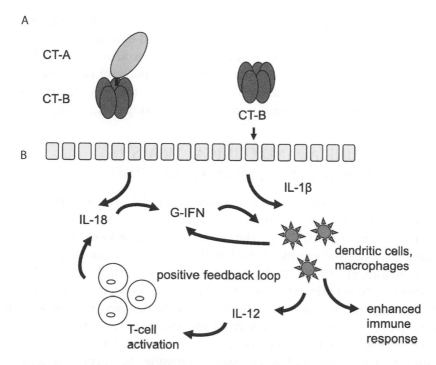

FIGURE 7.3 (A) Structure of CT holotoxin. Spatial orientation of subunits CT-A and CT-B pentamer are shown. (B) Model for CT-B mucosal adjuvanticity. Antigens that enter the GALT in the absence of adjuvant are presented to T cells by nonactivated antigen-presenting cells with low levels of costimulation. Adjuvants such as CT-B, however, can change this environment in several ways. In the GALT, CT-B first binds to intestinal epithelial cells, inducing IL-1β and IL-18 secretion, and as a result, gamma-IFN secretion from several types of immune response cells, including macrophages and dendritic cells. These cells induce IL-12 secretion, which acts synergistically with IL-18 to produce more gamma-IFN. The positive feedback loop that results increases the expansion of antigen-specific T cells and activated antigen-presenting cells. Antigen-presenting cells such as macrophages and dendritic cells then present the antigen in a context that results in T cell activation and proliferation, thus evoking a greater immune response.

mucosal immune response (Figure 7.3B). CT-B has been demonstrated to inhibit the intracellular processing of an antigen by macrophages, and to enhance the presentation of an antigen on the surface of MHC class II molecules when intracellular processing has occurred (Matousek et al., 1996).

7.10.4 SAPONINS

Saponin, derived from the bark of the quillaja (*Quillaja saponaria*) tree, is also used as an adjuvant. Saponins are nontoxic when administered orally, possibly due to their inability to cross the epithelial wall of the gastrointestinal tract. Saponins are routinely used in the food industry at concentrations much greater than those required for adjuvant activity, are stable at room temperature, and are easy to purify. Saponins have been employed as oral adjuvants in conjunction with plant-made measles virus vaccine proteins in preliminary studies using mice (Pickering et al., 2006).

7.11   PRECLINICAL AND CLINICAL TRIALS INVOLVING PLANT-DERIVED VACCINES

The mucosal immune response to vaccines can be greatly improved by delivering vaccine antigens to the intestinal mucosa via oral administration rather than parenteral injection. To date, the majority of clinical trials using plant-derived vaccines have largely focused on combating diseases that predominate in the third world. Diseases of livestock are also being addressed. Some clinical studies that have been conducted so far using plant-based vaccines are listed in Table 7.3. In this next section, several examples of responses to plant-derived vaccines found in clinical trials are discussed. In these trials, the possibility of eliciting an immune response either by consumption of plant tissue expressing the vaccine antigen or by direct injection of the purified antigen is examined. On the whole, the results are promising; in several cases, plant-derived vaccines worked as well as if not even better than traditional vaccines. One explanation for this is the protective effect of encapsulation of vaccine antigens by plant tissue eaten as food, which then enables the vaccine antigen to avoid degradation before it is presented to the Peyer's patches in the gut.

7.12   DIARRHEAL DISEASES

Preliminary trials were conducted in mice (Chikwamba et al., 2002; Lauterslager et al., 2001). Two of the most devastating diarrheal diseases are enterotoxigenic *E. coli* (ETEC) and Norwalk virus (NV). The B subunit of ETEC (LT-B) has been shown to bind to GM1 gangliosides of epithelial cells and is a potent stimulator of the mucosal immune response, but is not itself diarrheagenic. NV is composed of a single capsid protein that assembles spontaneously into virus-like particles (VLP) that stimulate the immune response. In a human clinical trial, transgenic potato or corn expressing either LT-B or NV antigens were fed to adult volunteers (Mason et al., 1996; Tacket et al., 2004, Tacket, 2007; Chikwamba et al., 2002; Lauterslager et al., 2001). Fourteen healthy adults ingested amounts of 50 or 100 g of raw transgenic or nontransformed potato that had been randomized in a double-blind fashion. A second and third dose were given on

TABLE 7.3
Examples of Clinical Studies Using Plant-Based Vaccines

Disease	Plant Used	Antisera Raised Against	Reference
Enterotoxigenic *E. coli*, ETEC	Potato, maize	LT-B	Tacket et al., 2004 Beyer et al., 2007
Norwalk virus	Potato, maize	NV	Mason et al., 1996
Hepatitis B virus	Potato	HBsAg	Kong et al., 2001
Rabies virus	Spinach	Spike protein	Modelska et al., 1998
Human papillomavirus	Potato, tobacco	L1 capsid protein	Warzecha et al., 2005 Biemelt et al., 2003 Kohl et al., 2006
Anthrax	Tobacco	Protective antigen (PA)	Koya et al., 2005
SARS	Tomato, tobacco	S protein	Pogrebnyak et al., 2005
Measles virus	Lettuce	MV-H protein	Webster et al., 2002, 2006
Swine transmissible, gastroenteritis virus	Maize	Spike protein	Streatfield et al., 2001 Lamphear et al., 2004
Staphylococcus aureus	Cowpea	D2 peptide of fibronectin-binding protein (FnBP)	Brennan et al., 1999
E. coli 0157:H7	Tobacco	Intimin protein	Judge et al., 2004
Strain K88 of enterotoxigenic *E. coli*	Tobacco	FaeG of K88 fimbrial antigen	Huang et al., 2003
Japanese cedar pollen allergens	Rice	Cry jI, Cry jII	Takagi et al., 2005
Foot and mouth disease virus	Alfalfa	VP1	Wigdorovitz et al., 1999
Respiratory syncytial virus	Tomato	F protein	Sandhu et al., 2000
Sunflower seed albumin	Narrow leaf lupin	SSA	Smart et al., 2003
Canine parvovirus	Tobacco chloroplasts	Epitope 2L21	Molina et al., 2005 *—continued*

TABLE 7.3 (continued)
Examples of Clinical Studies Using Plant-Based Vaccines

Disease	Plant Used	Antisera Raised Against	Reference
Bubonic plague (*Yersinia pestis*)	Tomato	F1-V	Alvarez et al., 2005
Cancer cells	Tobacco	Anti-Lewis Y MAb	Brodzik et al., 2006
Tetanus	Tobacco	Tetanus toxin Fragment C	Tregoning et al., 2005
Tuberculosis	*Arabidopsis*	ESAT-6	Rigano et al., 2005
E. coli	Potato	CFA/1 fibrial protein	Lee et al., 2004
Rabbit hemorrhagic disease virus	Potato	Capsid protein	Martin-Alonso et al., 2003
Influenza virus	Tobacco	HA protein	Shoji et al., 2008

days 7 and 21. In this study, antibody secreting cells were detected 7 days after ingestion. Volunteers who consumed potato- or corn-based LT-B vaccines also developed high increases in IgG; many of these developed fourfold rises in IgA anti-LT. LT neutralization assays were performed in Y-1 adrenal cells. Eight of eleven volunteers developed high neutralization titers (>1) of the individuals who ingested two to three doses of transgenic potatoes expressing NV CP; 95% developed significant rises in IgA titers. This study concluded that both humoral and systemic immune responses were elicited and that LT-B delivered in a potato cell could induce a similar amount of B cell priming as is determined from challenge by enterotoxigenic *E. coli* itself (Tacket, 2007).

More recently, a study was performed to identify the maximum non-immunostimulatory dose of transgenic maize expressing LT-B in mice. In this study, the level of LT-B that would not necessarily stimulate detectable antibody levels but would nonetheless result in the priming of memory cells for a later response was determined (Beyer et al., 2007). Serum IgA was found to be the most sensitive measure for detection of immunological priming and 0.002 μg was the highest dose of LT-B that could be administered using an intermittent feeding schedule without resulting in either an antibody response or immune priming. The authors were then able to use this along with additional data to estimate the dose that would be safe and nonimmunogenic for humans. This data will be useful for those who wish to provide doses of the vaccine protein to people for the greatest efficacy, while at the same time preventing the induction of oral tolerance.

7.13   HEPATITIS B VIRUS (HBV)

Infection by hepatitis B is among the principal causes of mortality in developing countries. Hepatitis B Virus surface antigen (HBsAg), is one of the most popular antigens selected for vaccine development. HBsAg has been demonstrated to form intact immunogenic virus-like particles in transgenic plant tissue. A mouse model has been used to compare the immunogenic responses to plant-derived and yeast-derived HBsAg (Kong et al., 2001). In this study, peeled potato tubers were fed to mice once a week for 3 weeks. Each feeding consisted of 5 g potato tuber (42 µg of HbsAg per dose). Mice fed transgenic tubers exhibited anti-HBsAg antibodies 1 week after the first two doses, peaked at 4 weeks after the third dose and returned to baseline levels at 11 weeks after the third dose. Control mice who were fed nontransgenic potato exhibited no elevated anti-HBsAg antibody response. In addition to this, no primary response was determined for mice that were fed yeast-derived HbsAg (Kong et al., 2001). The strong primary response exhibited by mice fed with potato-derived HbsAg is believed to be the result of protection by encapsulation. It is conceivable that digestion of the potato within the gut permitted the release of antigens near the Peyer's patches, resulting in a more robust immune response. The intact VLPs comprised of HBsAg that were visualized in these potatoes most likely render the antigen more immunogenic than the yeast-derived vaccine. To determine whether memory B cells had also been established as part of the immune response, mice who were first primed with potato derived HBsAg then received a parenteral boost of yeast-derived rHBsAg. These mice exhibited a strong secondary response that lasted for over 5 months (Kong et al., 2001).

In a later study, a double-blind, placebo controlled Phase 1 human clinical trial was performed. One hundred gram doses of uncooked transgenic potato tubers expressing approximately 8.5 µg/g HBsAg were fed to individual volunteers who had been previously vaccinated. Ten out of sixteen volunteers who ingested three doses of the transgenic potato tubers exhibited increased levels of serum anti-HBsAg titers, while none of the volunteers who ate the nontransformed potatoes provided as controls displayed a similar increase (Thanavala et al., 2005). The overall results of these studies present a solid foundation regarding the potential of plants to provide useful and much needed vaccines and other therapeutic agents to those in developing countries who have poor access to the benefits of modern medicine.

7.14   RABIES VIRUS

Rabies virus infection remains a significant global problem. A vaccine directed toward rabies virus spike protein has been developed in plants using a recombinant alfalfa mosaic virus (AlMV) vector and is discussed in detail in Chapter 3. This AlMV vector harboring the gene encoding the spike protein was then used to infect spinach (Modelska et al., 1998; Yusibov et al., 1999). There are several advantages to using spinach for expression of vaccine proteins. While the protein content of spinach is low, leaf material is amenable to freeze-drying, which enriches the antigen concentration on a per gram basis. Spinach that is freeze-dried does not require refrigeration, and as a result has a prolonged shelf-life. Spinach leaves containing virus, as well as virus purified from spinach, were used to immunize mice parenterally or orally. A protective antibody response was observed in mice after oral administration with purified virus. Mice fed with spinach leaves that contained the recombinant virus particles exhibited a higher level of immune response, suggesting that plant cells had a protective effect and enhanced virus particle delivery to the Peyer's patches. Orally immunized mice infected with an attenuated strain of rabies virus were able as a result to recover more rapidly than control mice, providing further evidence of the success of this vaccine.

7.15   HUMAN PAPILLOMAVIRUS (HPV)

Human papillomavirus is a major causative agent of cervical cancer in women in developing countries, discussed further in Chapter 2. To date, immunization studies using a plant-derived vaccine against human papillomavirus have been performed in mice. Biemelt et al. (2003) showed that 40 ng of either plant- or insect-derived virus-like particles (VLPs) composed of L1 of HPV were equally immunogenic. Furthermore, approximately half of mice fed transgenic tubers were shown to develop L1-specific antibodies, indicating that oral ingestion of L1-positive tubers could prime a humoral response against the L1 protein. In 2005, Warzecha et al. introduced a codon-optimized version of the L1 capsid protein of HPV into tobacco potato plants. Empty capsids were shown to form in these plants. Mice that consumed potato tubers expressing this codon optimized version of L1 elicited a significant enhanced serum antibody response after a "boost," indicating that plant-derived HPV can be considered a suitable vaccine by oral ingestion.

In addition to this, the potential of producing a plant-made vaccine against a papillomavirus using a virus-based vector has been explored. In this case, the L1 capsid protein of control rabbit papillomavirus (CRPV), a model system for the study of papillomavirus-host interactions, was inserted into a tobacco mosaic virus (TMV)-based vector. Rabbits inoculated with extracts of plants that either transiently express L1 in this TMV expression vector or that express L1 via conventional transgenic plants were both shown to be protected against challenge with high doses of the infectious virus (Kohl et al., 2006).

7.16 ANTHRAX

Anthrax is caused by *Bacillus anthracis*, Gram-positive spore-forming bacteria. The disease is acquired by inhalation or ingestion of spores. Anthrax is classified as a Category A biological warfare agent due to the rapidness and acuteness of the disease and its high mortality rate. One of the proteins expressed by anthrax, known as protective antigen (PA), is named for its ability upon immunization to elicit a protective immune response against anthrax. PA has been expressed at high levels in tobacco chloroplasts (see Chapter 3). Aziz et al. (2002) demonstrated that mature tobacco leaves accumulated levels as great as 14.2% of total soluble protein. A series of experiments were performed to determine the efficacy of this potential vaccine in mice. An in vitro macrophage lysis assay showed that chloroplast-derived PA was fully biologically active, at levels comparable to that of the *B. anthracis*-derived PA used as a positive control. Mice injected with chloroplast-derived PA exhibited an IgG response at titers comparable to *B. anthracis*-purified PA, suggesting that the plant-derived PA was correctly folded and functional. Sera taken from mice immunized with tobacco-derived PA at 15 days after the third immunization showed a similar ability to neutralize PA as compared to the *B. anthracis*-derived counterpart. Immunized mice were then challenged with and survived a lethal dose of anthrax LT (lethal toxin), further demonstrating the immunoprotective properties of chloroplast-derived PA (Koya et al., 2005).

7.17 SEVERE ACUTE RESPIRATORY SYNDROME (SARS)

In the past few years, there has been an increased demand for an effective vaccine against SARS (severe acute respiratory syndrome), the causative agent of which is a coronavirus. It has been determined that the spike protein (S protein) of the SARS coronavirus and its truncated fragments are the best candidates for vaccine development. In a recent study, Pogrebnyak et al. (2005) expressed the N-terminal fragment of the S protein (S1) at high levels in both tomato and tobacco plants. Mice that were fed lyophilized tomato fruit containing the antigen exhibited increased IgA levels in their feces. Significant titers of S protein-specific IgG were also detected in the sera of mice that were immunized parenterally and then boosted with S1 protein expressed in tobacco. A more detailed analysis revealed a high IgG1 immune response and significant IgG2a and IgG2b responses, but no changes in the levels of IgG3, IgM, IgA, or IgE. The results of this study suggest that a Th2-type response takes place in animals primed with plant-derived material expressing S1 of SARS coronavirus, as compared to those animals that are primed with purified antigen and elicit a Th1-type response.

7.18   MEASLES VIRUS

Measles is a highly contagious viral disease that is contracted through the respiratory tract. In resource-poor countries, the case fatality rate of measles is several hundred times that of developing nations. For example, in 2004, over 30 million cases of measles were reported. Eradication of the virus is made difficult by its highly contagious nature, and limitations of the current live MV vaccine, such as the requirement for refrigeration during transport and storage, as well as the syringes, medical infrastructure, etc., required for subcutaneous administration.

In a study conducted by Webster et al. (2002), high-titer MV-neutralizing antibodies were generated in mice by combining a plant-derived MV-H protein vaccine with a MV-H DNA vaccine in a prime-boost vaccination strategy. Mice were given an intramuscular dose of either MV-H or control DNA, followed by oral administration of MV-H or plant extract. Over 90% of mice exhibited an IgG response. Those mice boosted with plant-derived MV-H had a much greater IgG response (a response dominated by IgG1 and mediated by the Th2 pathway) than mice boosted with control plant extract. The results of this study also show that the administration of a DNA vaccine followed by a plant-derived antigen booster may represent a feasible strategy to evoke an immune response for measles as well as for other infectious diseases.

In a more recent study, Webster et al. (2006) report the expression and characterization of lettuce-derived measles vaccine. The MV-H protein expressed in lettuce was demonstrated to be immunogenic in mice following intraperitoneal injection in the absence of adjuvant in addition to intranasal inoculation in the presence of a mucosal adjuvant. The highest response was observed in mice primed first with MV-H DNA and then boosted with an oral formulation of freeze-dried MV-H lettuce in conjunction with a mucosal adjuvant. In addition to this, the type of immune response was found to depend largely on the manner in which MV-H is presented to the immune system. Secreted and soluble forms of MV-H were demonstrated to induce a Th2 type response, while membrane-bound MV-H protein was found to be associated with a Th1 response.

7.19 INFLUENZA VIRUS

Influenza virus is a respiratory pathogen that is responsible for a high degree of yearly mortality around the globe. The fear of the avian bird flu has heightened the need for new, inexpensive vaccines that can be readily administered across the globe. Recently, Shoji et al. (2008) have generated the full length hemagglutinin protein from the H3N2 strain of the virus in transgenic plants. Mice immunized with this plant-derived HA exhibited both humoral (IgG1, IgG2a, and IgG2b) and cell-mediated (interleukin-5, γ-interferon) responses, respectively. Antibody titers obtained were sufficient to neutralize the virus and inhibit serum hemagglutination, indicating that plant-derived HA can evoke an immune response that is both effective and protective.

7.20 SWINE-TRANSMISSIBLE GASTROENTERITIS VIRUS

One of the most prevalent diseases being addressed in veterinary medicine today is that of swine transmissible gastroenteritis virus (STGV). The company ProdiGene has succeeded in expressing the spike protein of swine transmissible gastroenteritis virus in maize grain. Oral delivery studies have been conducted using both piglets and sows (Streatfield et al., 2001). Piglets who were first primed then administered the plant-derived vaccine elicited a strong immune response and were protected against STGV. Assessment of an oral booster application in sows who were administered first with the live modified virus vaccine and later maize expressing the vaccine protein exhibited a stimulation in serum, colostrum and early milk antibodies at levels comparable to boosting via conventional vaccines (Streatfield et al., 2001; Lamphear et al., 2004).

7.21 CANINE PARVOVIRUS

Canine parvovirus, causing acute gastroenteritis and myocarditis in dogs, consists of three different capsid proteins (VP1, VP2, and VP3). The amino terminus of VP2 encodes the 21-amino-acid long B cell 2L21 epitope. A CT-B:2L21 fusion protein was constructed in transgenic tobacco chloroplasts and was previously shown to be expressed at high levels (Chapter 3). Combined immunizations were conducted using either leaf extracts or pulverized tissues from transgenic plants (Molina et al., 2005). Upon oral delivery, pulverized tissues were able to induce both IgG and IgA antibody responses in mice and rabbit animal models. Parenteral administration followed by oral delivery also resulted in the induction of anti-2L21 antibodies.

7.22   ORAL TOLERANCE TO ANTIGENS

Oral tolerance, as discussed earlier in this chapter, is considered to be the induction of a state of systemic unresponsiveness to a particular administered antigen. Takagi et al. (2005) developed transgenic rice plants whose seed accumulated mouse T cell epitope peptides specific for pollen allergens of *Cryptomoeria japonica* (Japanese cedar). The T cell epitope peptides corresponding to Cry jI and Cry jII pollen antigens were expressed together with soybean storage protein glycinin AlaB1b as part of a fusion protein. Oral consumption of transgenic rice by mice prior to systemic challenge with pollen resulted in allergen-induced oral tolerance, accompanied by a dramatic inhibition of sneezing. The systemic unresponsiveness corresponded with a reduction of pollen allergen-specific Th2-mediated IgE responses and histamine release, while the CD4+ T cell-proliferative response remained unaffected. This represents a proof-of-concept study of the ability of oral tolerance to be induced by plant-derived antigens.

7.23   CONCLUSIONS

The results of these preliminary clinical trials support the potential of plants to become oral delivery vehicles for vaccines. Those who ingest plant cells containing vaccine antigen exhibit heightened levels of immune response, as well as increased protection, and recover more rapidly from disease than control human volunteers or animals. The results of the studies presented here hold great promise for the use of plant-derived vaccines in the future. Indeed, the heterologous prime-boost vaccination strategy in which a vaccine protein is delivered in two different formats as incorporated in several of the studies listed in this chapter has demonstrated the versatility for use of plant-derived vaccines in inducing an immune response in both human volunteers as well as in animal models. In this strategy, the first vaccine is used to prime the antigen-specific T cells, then a different boosting vaccine targeting the same antigen is used to induce an expansion of antigen-specific memory cells. This prime-boost strategy has resulted in an increase in antigen-specific CD4+ and CD8+ T cells, and has resulted in improved protection against challenge by the pathogen in question (Chalmers, 2006).

The provocation of mucosal immunity against a given antigen can be achieved by other means besides oral ingestion. For example, intranasal administration of vaccine proteins can improve local mucosal immunity and enable large populations to be immunized at a lower cost. Plant-derived vaccines provide hope for more immunogenic, more effective and less expensive vaccination strategies against both respiratory (e.g., anthrax) as well as GI tract pathogens (e.g., ETEC) in the not so distant future.

REFERENCES

Alvarez, M.L., Pinyerd, H.L, Crisantes, J.D., Rigano, M.M., Pinkhasov, J., Walmsley, A.M., Mason, H.S., and Cardineau, G.A. (2005). Plant-made subunit vaccine against pneumonic and bubonic plague is orally immunogenic in mice. *Vaccine* 24(14): 2477–2490.

Aziz, M.A., Singh, S., Kumar, P.A., and Bhatnagar, R. (2002). Expression of protective antigen in transgenic plants: a step towards edible vaccine against anthrax. *Biochem. Biophys. Res. Commun.* 299: 345–351.

Beyer, A., Wang, K., Umble, A.N., Wolt, J.D., and Cunnick, J.E. (2007). Low-dose exposure and immunogenicity of transgenic maize expressing the Escherichia coli heat-labile toxin B subunit. *Environ. Health Perspect.* 115(3): 354–360.

Biemelt, S., Sonnewald, U., Galmbacher, P., Willmitzer, L., and Muller, M. (2003). Production of human papillomavirus type 16 virus-like particles in transgenic plants. *J. Virol.* 77(17): 9211–9220.

Botturi, K., Vervloet, D., and Magnan, A. (2007). T cells and allergens relationships: are they that specific? *Clin Exp. Allergy.* 37(8): 1121–1123.

Brennan, F.R., Bellaby, T., Helliwell, S.M., Jones, T.D., Kamstrup, S., Dalsgaard, K., Flock, J.-I., and Hamilton, D.O. (1999). Chimeric plant virus particles administered nasally or orally induce systemic and mucosal immune responses in mice. *J. Virol.* 73(2): 930–938.

Brodzik, Glogowska, M., Bandurska, K., Okulios, M., Deka, D., Ko, K., van der Linden, J., Leusen, J.H.W., Pogrebnyak, N., Golovkin, M., Steplewski, Z., and Koprowski, H. (2006). Plant-derived anti-Lewis Y MAb exhibits biological activities for efficient immunotherapy against human cancer cells. *Proc. Natl. Acad. Sci. U.S.A.* 103(923): 8804–8809.

Chalmers, W.S.K. (2006). Overview of new vaccines and technologies. *Veterinary Microbiol.* 117: 25–31.

Chikwamba, R., Cunnick, J., Hathaway, D., McMurray, J., Mason, H., and Wang, K. (2002). A functional antigen in a practical crop: LT-B producing maize protects mice against *Escherichia coli* heat labile enterotoxin (LT) and cholera toxin (CT). *Transgenic Res.* 11(5): 479–493.

Clavel, T. and Haller, D. (2007). Molecular interactions between bacteria, the epithelium, and the mucosal immune system in the intestinal tract: implications for chronic inflammation. *Curr. Issues Intestinal Microbiol.* 8(2): 25–43.

Corthesy, B. (2007). Roundtrip ticket for secretory IgA: role in mucosal homeostasis? *J. Immunol.* 178: 27–32.

Daniell, H., Lee, S.-B., Panchal, T., and Wiebe, P.O. (2001). Expression of the native cholera toxin B subunit gene and assembly as functional oligomers in transgenic tobacco chloroplasts *J. Mol. Biol.* 311: 1001–1009.

Faria, A.M. and Weiner, H.L. (2006). Oral tolerance: therapeutic implications for autoimmune diseases. *Clin. Dev. Immunol.* 13(2–4): 143–157.

Fazilleau, N., McHeyzer-Williams, L.J., and McHeyzer-Williams M.G. (2007). Local development of effector and memory T helper cells. *Curr. Opin. Immunol.* 19(3), 259–267. Epub 2007 Apr 8.

Foss, D.L. and Murtaugh, M.P. (2000). Mechanisms of vaccine adjuvanticity at mucosal surfaces. *Anim. Health Res. Rev.* 1(1): 3–24.

Hajishengallis, G., Arce, S., Gockel, C.M., Connell, T.D., and Russell, M.W. (2005). Immunomodulation with enterotoxins for the generation of secretory immunity or tolerance: applications for oral infections. *J. Dental Res.* 84(12): 1104–1116.

Huang, Y., Liang, W., Pan, A., Zhou, Z., Huang, C., Chern, J., and Zhang, D. (2003). Production of FaeG, the major subunit of K88 fimbriae, in transgenic tobacco plants and its immunogenicity in mice. *Infect. Immun.* 71(9): 5436–5439.

Janeway, C.A. et al. (eds.). (1999). *Immunobiology. The Immune System in Health and Disease*, 4th ed., New York: Garland.

Judge, N.A., Mason, H.S., and O'Brien, A.D. (2004). Plant cell-based intimin vaccine given orally to mice primed with intimin reduces times of *Escherichia coli* O157:H7 shedding in feces. *Infect. Immun.* 72(1): 168–175.

Kang, W. and Kudsk, K.A. (2007). Is there evidence that the gut contributes to mucosal immunity in humans? *J. Parenter. Gen. Nutr.* 31(3): 246–258.

Kohl, T., Hitzweoth, I.I., Stewart, D., Varsani, A., Govan, V.A., Christensen, N.D., Williamson, A.-L., and Rybicki, E.P. (2006). Plant-produced cottontail rabbit papillomavirus L1 protein protects against tumor challenge: a proof of concept study. *Clin. Vaccine Immunol.* 13(8): 845–853.

Kong, Q., Richter, L., Yang, Y.F., Arntzen, C., Mason, H.S., and Thanavala, Y. (2001). Oral administration with hepatitis B surface antigen expressed in transgenic plants. *Proc. Natl. Acad. Sci. U.S.A.* 98(20): 11539–11544.

Koya, V., Moayeri, M., Leppin, S.H., and Daniell, H. (2005). Plant-based vaccine: mice immunized with chloroplast-derived anthrax protective antigen survive anthrax lethal toxin challenge. *Infect. Immun.* 73(12): 8266–8274.

Lamphear, B.J., Jilka, J.M., Kesl, L., Welt, M., Howard, J.A., and Streatfield, S.J. (2004). A corn-based delivery system for animal vaccines: an oral transmissible gastroenteritis virus vaccine boosts lactogenic immunity in swine. *Vaccine* 22(19): 2420–2424.

Lauterslager, T.G., Florack, D.E., van der Wal, T.J., Molthoff, J.W., Langeveld, J.P., Bosch, D., Boersma, W.J., Hilgers, L.A. (2001). Oral immunisation of naive and primed animals with transgenic potato tubers expressing LT-B. *Vaccine* 19: 2749–2755.

Lauterslager, T.G.M. and Hilgers, L.A.T. (2002). Efficacy of oral administration and oral intake of edible vaccines. *Immunol. Lett.* 84: 183–190.

Lee, J.-Y., Yu, J., Henderson, D., Langridge, W.H.R. (2004). Plant-synthesized *E. coli* CFA/I fibrial protein protects Caco-2 cells from bacterial attachment. Vaccine 23(2): 222–231.

Limaye, A., Koya, V., Samsam, M., and Daniell, H. (2006). Receptor-mediated oral delivery of a bioencapsulated green fluorescent protein expressed in transgenic chloroplasts into the mouse circulatory system. *FASEB J.* 20(7): 959–961.

Lu, B. (2006). The molecular mechanisms that control function and death of effector CD4+ T cells. *Immunol. Res.* 36(1–3): 275–282.

Lu, D. and Hickey, A.J. (2007). Pulmonary vaccine delivery. *Expert Rev. Vaccines* 6(2): 213–226.

Macperson, A.J. and Uhr, T. (2004) Compartmentalization of the mucosal immune responses to commensal intestinal bacteria. *Ann. NY Acad. Sci.* 1029: 36–43.

Mason, H.S., Ball, J.M., Jian-Jian Shi, Jiang, X., Estes, M.K., and Arntzen, C.J. (1996). Expression of Norwalk virus capsid protein in transgenic tobacco and potato and its oral immunogenicity in mice. *Proc. Natl. Acad. Sci. U. S. A.* 93: 5335–5340.

Martin-Alonso, J.M., Castanon, S., Alondso, P., Patra, F., and Ordas, R. (2003). Oral immunization using tuber extracts from transgenic potato plants expressing rabbit hemorrhagic disease virus capsid protein. *Transgenic Res.* 12: 127–130.

Modelska, A., Dietzschold, B., Sleysh, N., Fu, Z.F., Steplewski, K., Hoopwer, D.C., Koprowski, H., and Yusibov, V. (1998). Immunization against rabies with plant-derived antigen. *Proc. Natl. Acad. Sci. U.S.A.* 95: 2481–2485.

Molina, A., Veramendi, J., and Hervas-Stubbs, S. (2005). Induction of neutralizing antibodies by a tobacco chloroplast-derived vaccine based on a B-cell epitope from canine parvovirus. *Virology* 342 (2): 266–275.

Montufar-Sulis, D., Garza, T., and Klein, J.R. (2007). T-cell activation in the intestinal mucosa. *Immunol. Rev.* 215: 189–201.

Ogra, P., Faden, H., and Welliver, R.C. (2001) Vaccination strategies for mucosal immune responses. *Clin. Microbiol. Rev.* 14(2): 430–445.

Parham, P. (2005). *The Immune System*, 2nd ed. New York: Garland Science.

Pickering, R.J., Smith, S.D., Strugnell, R.A., Wesselingh, S.L., and Webster, D.E. (2006). Crude saponins improve the immune response to an oral plant-made measles vaccine. *Vaccine* 24: 144–150.

Pogrebnyak, N., Golovkin, M., Andrianov, V., Spitsin, S., Smirnov, Y., Egolf, R., and Koprowski, H. (2005). Severe acute respiratory syndrome (SARS) S protein production in plants: development of recombinant vaccine. *Proc. Natl. Acad. Sci. U.S.A.* 102(25): 9062–9067.

Prlic, M. and Bevan, M.J. (2006). An antibody paradox, resolved. *Science* 311(5769):1875–1876.

Reiner, S.L. (2007). Development in motion: helper T cells at work. *Cell* 6,129(1): 33–36.

Rigano, M.M., Dreitz, S., Kipnis, A.-P., Izzo, A.A., and Walmsley, A.M. (2005). Oral immunogenicity of a plant-made subunit tuberculosis vaccine. *Vaccine* 24(5): 691–695.

Roitt, I. et al. (eds.). (2002). *Immunology*, 5th ed. London: Mosby.

Rojas, R. and Apodaca, G. (2002). Immunoglobulin transport across polarized epithelial cells. *Nat. Rev. Mol. Cell Biol.* 3: 1–12.

Sandhu, J.S., Krasnyanski, S.F., Domier, L.L., Korban, S.S., Osadjan, M.D., and Buetow, D.E. (2000). Oral immunization of mice with transgenic tomato fruit expressing respiratory syncytial virus-F protein induces a systemic immune response. *Transgenic Res.* 9(2): 127–135.

Shoji, Y., Chichester, J.A., Bi, H., Musiychuk, K., de la Rosa, P., Goldschmidt, L., Horsey, A., Ugulava, N., Palmer, G.A., Mett, V., and Yusibov, V. (2008). Plant-expressed HA as a seasonal influenza vaccine candidate. *Vaccine* 26(23): 2930–2934.

Smart, V., Foster, P.S., Rothenbwerg, M.E., Higgens, T.J., and Hogan, S.P. (2003). A plant-based allergy vaccine suppresses experimental asthma via an IFN-gamma and CD4+CD45Rblow T cell-dependent mechanism. *J. Immunol.* 171(4): 2116–2126.

Streatfield, S.J. (2005a). Mucosal immunization using recombinant plant-based oral vaccines. *Methods.* 38(2): 150–157.

Streatfield, S.J. (2006). Delivery of plant-derived vaccines. *Exp. Opin. Drug Delivery* 2(4): 719–728.

Streatfield, S.J., Jilka, J.M., Hood, E.E., Turner, D.D., Bailey, M.R., Mayor, J.M., Woodard, S.L., Beifuss, K.K., Horn, M.E., Delaney, D.E., Tizard, I.R., and Howard, J.A. (2001). Plant-based vaccines: unique advantages. *Vaccine* 19: 2742–2748.

Tacket, C.O. (2007). Plant-based vaccines against diarrheal diseases. *Trans. Am. Climatol. Assoc.* 118: 79–87.

Tacket, C.O. (2009). Plant-based oral vaccines: results of human trials. *Curr. Top. Micro. Immunol.* 332: 103-117.

Tacket, C.O., Passeti, M.F., Edelman, R.E., Howad, J.A., and Streeatfield, S.J. (2004). Immunogenicity of recombinant LT-B delivered orally to humans in transgenic corn. *Vaccine* 22: 4385–4389.

Takagi, H., Hiroi, T., Yang, L., Tada, Y., Yuki, Y., Takamura, K., Ishimitsu, R., Kawauchi, H., and Takaiwa, F. (2005). A rice-based edible vaccine expressing multiple T cell epitopes induces oral tolerance for inhibition of Th2-mediated IgE responses. *PNAS* 102(48): 17525–17530.

Thanavala, Y., Mahoney, M., Pal, S., Scott, A., Richter, L., Natarajan, N., Goodwin, P., Arntzen, C.J., and Mason, H.S. (2005). Immunogenicity in humans of an edible vaccine for hepatitis B. *Proc. Natl. Acad. Sci. U.S.A.* 102(9): 3378–3382.

Thanavala, Y, Huang, Z., and Mason, H. (2006). Plant-derived vaccines: a look back at the highlights and a view to the challenges on the road ahead. Exp. Rev. *Vaccines* 5(2): 249-260.

Tregoning, J.S., Clare, S., Bowe, F., Edwards, L., Fairweather, N., Qazi, O., Nixon, P.J., Maliga, G., Dougan, G., Hussell, T. (2005). Protection against tetanus toxin using a plant-based vaccine. *Eur. J. Immunol.* 35: 1320–1326.

Twigg, H.L. (2005). Humoral immune defence (antibodies). *Proc. Am. Thoracic Soc.* 2: 417–421.

Vajdy, M. (2006). Generation and maintenance of mucosal memory B cell responses? *Curr. Med. Chem.* 13(25): 3023–3037.

Walker, R.I. (1994). New strategies for using mucosal vaccination to achieve more effective immunization. *Vaccine* 12(5): 387–400.

Warzecha, H., Mason, H.S., Lane, C., Tryggvesson, A., Rybicki. E., Williamson, A.-L., Clements, J.D., and Rose, R.C. (2005). Oral immunogenicity of human papillomavirus-like particles expressed in potato. *J. Virol.* 77(16): 8702–8711.

Webster, D.E., Cooney, M.L., Huang, Z., Drew, D.R., Ramshaw, I.A., Dry, I.B., Strugnell, R.A., Martinm, J.L., and Wesselingh, S. L. (2002). Successful boosting of a DNA measles immunization with an oral plant-derived measles virus vaccine. *J. Virol.* 76(15): 7910–7912.

Webster, D.E., Smith, S.D., Pickering, R.J., Strugnell, R.A., Dry, I.B., and Wesselingh, S.L. (2006). Measles virus hemagglutinin protein expressed in transgenic lettuce induces neutralising antibodies in mice following mucosal vaccination. *Vaccine* 24(17): 3544–3548.

Wigdorovitz, A., Carrillo, C., Dus Santos, M.J., Trono, K., Peralta, A., Gomez, M.C., Rios, R.D., Franzone, P.M., Sadir, A.M., Escribano, J.M., and Borca, M.V. (1999) Induction of a protective antibody response to foot and mouth disease virus in mice following oral or parenteral immunization with alfalfa transgenic plants expressing the structural protein VP1. *Virology* 255: 347–353.

Woodfolk, J.A. (2007). T-cell responses to allergens. *J. Allergy Clin. Immunol.* 119(2): 280–294.

Yusibov, V., Shivprasad, S., Turpen, T.H., Dawson, W., and Koprowski, H. (1999). Plant viral vectors based on tobamoviruses. *Curr. Top. Microbiol. Immunol.* 240: 81–94.

8 Risk Analysis and Safety of Plant-Made Biopharmaceuticals

8.1 INTRODUCTION

Risk analysis can been defined as a means by which scientific, social, cultural, economic, and political issues that together form a consensus approach in public policy decisions in a particular discipline, such as agricultural biotechnology, can be addressed. In the case of science-based risk analysis, the potential risks must be assessed at each stage of the production process, from research and development to commercialization, and strategies must be developed to deal with them. These strategies can then be brought in front of political decision makers who must contend with the public perception of risk. With respect to plant-made biopharmaceuticals, the overall amount of risk is the measure used by the FDA to determine whether this approval should be granted (Shama and Peterson, 2004; Farrow, 2004).

Risk analysis has been proven to retain the flexibility necessary to make it a useful model system for addressing the countless issues that are found to be associated with plant-derived pharmaceuticals (Wolt and Peterson, 2000). Over the past few years, a great deal of information and experience has steadily accumulated with respect to risk analysis of pharmaceuticals that are currently produced in bacterial and animal cell bioreactor systems. Risk analysis has also been performed on transgenic crops used for food production as well as for other applications. As a result, elements from each of these disciplines can be incorporated into the design of optimal production and testing policies and practices. Risk analysis has been employed to cover a series of important issues regarding the large-scale manufacture of plant-made biopharmaceuticals, and will continue to present serious issues for researchers in the academic, corporate, and public health arenas to address (Miele, 1997; Ciliberti and Molinelli, 2005).

A few plant-derived biopharmaceutical products have now reached advanced clinical trials, and the regulatory process has been developed concurrently. There are few commercial plant-based pharmaceuticals currently on the market. The regulatory process follows the existing regulatory framework for the approval of GM plants and biopharmaceuticals and will be described in more detail below.

A number of potential risks are associated with plant-based pharmaceuticals; these include allergen exposure to the public, pollen transfer to wild species, nontarget organism exposure due to persistence of genetically engineered material in the environment, interspecies gene flow, and contamination of nontransgenic crops intended for human consumption. The role of risk analysis with respect to the impact of plant-based biopharmaceuticals on human health and the environment are discussed in this chapter.

8.2   RISK ANALYSIS AND PLANT-BASED BIOPHARMACEUTICALS

Since risk analysis plays an important role in public policy decision making, efforts have been made to devise a means by which to identify, control, and communicate the risks imposed by agricultural biotechnology. A paradigm of environmental risk assessment was first introduced in the United States by Peterson and Arntzen in 2004. In this risk assessment, a number of assumptions and uncertainties were considered and presented. These include (1) problem formulation, (2) hazard identification, (3) dose-response relationships, (4) exposure assessment, and (5) risk characterization. Risk assessment of plant-made pharmaceuticals must be reviewed on a case-by-case basis because the plants used to produce proteins each have different risks associated with them. Many plant-derived biopharmaceuticals will challenge our ability to define an environmental hazard (Howard and Donnelly, 2004). For example, the expression of a bovine-specific antigen produced in a potato plant and used orally in veterinary medicine would have a dramatically different set of criteria for assessment of risk than, as another example, the expression of a neutralizing nonspecific oral antibody developed in maize to suppress *Campylobacter jejuni* in chickens (Peterson and Arntzen, 2004; Kirk et al., 2005).

It is paramount that the knowledge produced from risk analysis be used effectively to deal with public perception. As we enter an ever-increasing technologically complex world, the general public feels more and more obligated to blindly trust scientists and public policy makers prior to fully understanding the basis of the new technology. A most obvious example of this has been demonstrated in the lack of a reliable knowledge base in Europe which ultimately resulted in low public acceptance of agricultural biotechnology (Peterson, 2000). The broad differences in public perception between Europe and North America can be attributed to dissimilar levels of knowledge and, as a consequence, trust. While Europeans and Americans together share only a limited understanding of many of the concepts underlying agricultural biotechnology, public leaders and policy makers in the United States have taken a more aggressive role in educating themselves with respect to its risks and benefits. Since these public policy makers are trusted in the United States, the technologies have been accepted more readily and risk perception in North America continues to be shaped by education based on sound scientific principles. Furthermore, a series of safeguards have been developed to ensure public confidence in quality and safety of the food supply. These safeguards have been set in place currently with the introduction of a new "bioeconomy"; that is, the maturation of agricultural biotechnology. In contrast, European public officials and scientists tend to be viewed as less trustworthy, resulting in a trend toward a lower public acceptance of innovative agricultural biotechnology by the general public (Peterson, 2000). The response of policy makers to such public stubbornness with regard to biotechnology has so far been to be slow or even fail to make decisions on public policy toward agricultural biotechnology. This explains Europe's state of affairs with respect to the many restrictions now in place concerning the generation of genetically modified crops within the European Union.

8.3 REGULATION OF PLANT-MADE BIOPHARMACEUTICALS

Due to the many issues concerning public perception and risk assessment described above, the approval and regulation of plant-made biopharmaceuticals by the U.S. government is even more extensive than the manner in which traditional pharmaceuticals are currently overseen. It is likely that more anticipated issues will come forward as products develop. Each problem is unique, and may have to be resolved on a case-by-case basis. A template of federal laws, agencies, and regulations govern the evaluation, production, and distribution of final biopharmaceutical products derived from recombinant plants. Such products fall under the authority of the FDA (U.S. Food and Drug Administration), the USDA (U.S. Department of Agriculture), and the EPA (Environmental Protection Agency), depending on the nature of the product and its intended use.

A regulatory framework and approval process was set up to avoid inadvertent release of material, to ensure environmental safety, and to protect the integrity of the plant product intended for food or animal feed. Much overlap exists between the agencies responsible for addressing these issues.

1. The FDA regulates biologics and drugs intended for use in humans.
2. The USDA regulates the addition of new plants and plant products into the environment. Specifically, genetically engineered plants require a permit or notification obtained from the USDA.
3. The EPA is involved in the regulation process if the transgenic plant expresses a pest- or herbicide-resistant engineered trait in addition to a biopharmaceutical.

FDA approval of plant-derived biopharmaceuticals requires the submission of a Biological License Application (BLA) or a New Drug Application (NDA). These applications request information that enables an assessment of the safety and efficacy of the product. In order to generate data for these documents, an IND (investigational new drug) application must be submitted as for all biologics, whether produced in plants or elsewhere. USDA approval is also required for all field testing of plant-based biopharmaceuticals. In this case, regulations are based on the principle of minimizing environmental impact by confinement and maximizing nontarget organism safety.

A number of documents have been issued by the FDA and are known as Points to Consider. These documents, generated by the Center for Biologics Evaluation and Research (CBER), can be updated and provide a flexible regulatory approach to novel issues in rapidly evolving research areas (www.fda.gov/cber/gdlns/bioplant.pdf.). Biopharmaceuticals derived from plants grow under highly regulated conditions in confined growing areas, strictly controlled by the USDA Animal and Plant Health Inspection Service (APHIS), and the FDA. In spite of some criticism, food crops remain the crop of choice for plant-made biopharmaceuticals because of the extensive agricultural knowledge and familiarity with these plants (Jaffe, 2004). A vast understanding of the genetics, agronomics, and the environmental impact of a particular plant remains critical for the development of management and containment strategies for plant-derived biopharmaceuticals (www.BIO.com).

Plant-made pharmaceutical research and development to the commercialization stage are reviewed and regulated by the USDA. This includes APHIS, which regulates the movement, importation, and field testing of plant-made pharmaceuticals through permitting and notification procedures; and the USDA Center for Veterinary Biologics (CVB), which regulates biological products from research and development to commercial distribution. FDA agencies include the Center for Biologics Evaluation and Research (CBER), and the Center for Drug Evaluation and Research (CDER), centers of the FDA that regulate biologic products for use in humans; the center for Food Safety and Applied Nutrition (CFSAN), the center at the FDA responsible for promoting and protecting the public's health by ensuring that the nation's food supply is safe, sanitary, and wholesome; and the Center for Veterinary Medicine (CVM), the center at the FDA that provides premarket consultation on food and feed safety. Regulations regarding development, testing, production, transportation, and commercialization of plant-made biopharmaceuticals are expected to adapt as the technology advances. Currently, a USDA field permit is required to plant crops that produce plant-made pharmaceuticals, and permits are granted on a case-by-case basis. All field trials are inspected by APHIS. For example, the number of permit applications submitted to the USDA starting in May 2003, was sixteen; the geographical range of the applications awarded is extensive (Jaffe, 2004). An example of the typical pathway of steps involved in generating plants that express pharmaceuticals and therapeutic proteins is provided in Figure 8.1.

8.4 QUALITY CONTROL AND THE MANUFACTURING OF THE PRODUCT

Quality control of plant-made biopharmaceuticals for commercial use includes the development of a seed bank based on the transgenic plant expressing the therapeutic product, the development of a population of plants grown from a seed bank, the harvesting of these plants, and their subsequent subjection to an extraction and purification process similar to

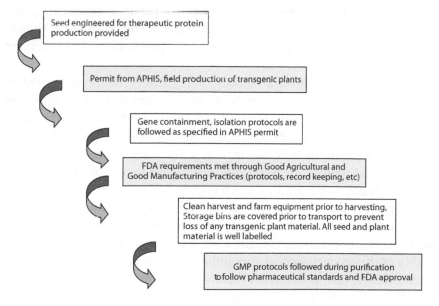

FIGURE 8.1 Typical pathway of regulatory steps involved in generating plants that express biopharmaceuticals.

traditional processes with cells or microorganisms. Each of these stages must be addressed independently to assure consistency and high quality of the final product.

In choosing the host plant type and tissue for producing plant-based biopharmaceuticals, concerns regarding the presence of potentially toxic or allergenic substances must be addressed. Issues such as confinement, potential unfavorable public opinion regarding the production system, potential routes of exposure for different plant hosts, and criteria used for choosing a host plant have resulted in the recommendation of certain characterization data for transgenic plants including copy number, vector used, transformation methods, stability of inserted DNA, etc.

There are many advantages to the manufacture of biopharmaceuticals in a plant-based system. In the case of vaccine and therapeutic proteins produced in plants, production costs are significantly lower than for animal and prokaryotic cell-based production systems. Since there is no need for skilled personnel to run the equipment or fermentors, production for plant-derived proteins has been estimated to be 2%–10% of the cost of microbial fermentation systems and 0.1% of the cost of mammalian cell cultures. In addition to this, transgenic animals and fermentation systems have limited scalability, and scale-up or scale-down can also be a slow and expensive process. On the other hand, the cultivation of transgenic plants for protein products is both inexpensive and fast. Production of vaccine and human therapeutic proteins in plants mitigates the hurdle of potential contamination by biohazardous agents such as viruses or prions, a significant problem in animal cell culture systems. Many useful technologies have already been developed and are currently in use for the purification of recombinant proteins from plant tissue.

While this is attractive, a number of regulatory issues connected to all potential pharmaceutical applications of plant-derived vaccine and therapeutic proteins remain (Howard and Donnelly, 2004). These general principles form the basis for the review of their manufacture and clinical testing. Biological products, whether they are derived from plant, bacterial, or animal sources, are heat labile, are subject to microbial contamination, can become damaged by mechanical shearing, and can potentially be inappropriately immunogenic or allergenic. Since plant material used in the purification process will likely be generated in a nonsterile environment, exposure to various contaminants that would affect the purity and quality of the product, such as agricultural chemicals, weather, insects and other wildlife, dirt, pollen from other plants, fungi, and bacteria, can be a problem. For this reason, the molecular features of plant-based biopharmaceuticals should be examined and compared, whenever possible, to their bacterial or animal-derived counterparts. Several concerns regarding quality control in the manufacturing of plant-based biopharmaceuticals are listed below.

8.4.1 CHARACTERIZATION OF THE PRODUCT

A number of concerns must be addressed with respect to the plant-based biopharmaceutical product. These include the inclusion of different impurities in the plant-derived material that are not present in other cell culture systems; the use and residual presence of pesticides, herbicides, and fungicides; the possibility of infestation by insect or fungal contaminants which have not been removed during the harvesting process; the presence of potential impurities from the host plant source such as biologically active plant-derived metabolites or alkaloids (e.g., nicotine); the presence of metals that could affect the safety of the plant-based product by inducing a toxic effect; and the presence of contaminant plant proteins or macromolecules which could affect product stability, allergenicity, and immunogenicity. Therefore, the whole manufacturing process, from qualification and storage of raw materials to sterile filling of containers for shipping, should be monitored for sterility. In addition, testing for the presence of toxins would have to be conducted at appropriate stages during the purification process. Good Laboratory Practice and Good Manufacturing Practice are critical for fully characterized, contaminant-free materials and appropriate quality assurance for reproducible results (Goldstein and Thomas, 2004).

8.4.2 PROTEIN INSTABILITY

Posttranslational modifications of plant-derived biopharmaceuticals have the potential to differ from those encountered in an animal or prokaryotic cell culture production system. As a result, relatively minor variations in production, purification, storage conditions, or formulation may cause large differences in biological activity or immunogenicity by affecting the fraction of correctly folded molecules in the final product. In addition, determination of the exact molecular structure with respect to protein folding and posttranslational modifications is not always possible. All of these characteristics raise regulatory issues that may differ from those traditionally associated with drug development. For example, changes in the glycosylation pattern of the Fc region of a monoclonal antibody molecule can affect receptor binding, antibody-mediated cytotoxicity, and proteolytic degradation. Differences in glycosylation patterns have been shown to affect the pharmacokinetics and activity of several cytokines and hormones that are naturally glycosylated in humans. Alternatively, a patient may develop an immune response to plant-derived oligosaccharide moieties present on a plant-based medicinal product. The consideration of a possible allergic reaction upon repeated administration must be taken into account in the design of appropriate clinical trials. In some cases, it may be appropriate to enzymatically remove unwanted glycol groups or use recombinant plant strains with altered glycosylation pathways for production of some therapeutic proteins for human consumption. These concerns are addressed in more detail in Chapter 5.

8.4.3 Genetic Stability

As it is imperative that the plant-derived biopharmaceutical product must be obtained repeatedly and on a consistent basis, a master cell culture bank, seed bank for transgenic plants, or virus seed stock for transient expression systems must be constantly maintained. Storage conditions must therefore be optimized to prevent contamination and ensure viability. Both transgene stability (e.g., reversion to wild type or sequence drift of plant virus expression vectors) and protein expression levels must be monitored in a representative plant of a given bank or stock to minimize any possible variation in expression levels that may affect safety and consistency of the final product. A program that monitors lot-to-lot consistency of the biochemical and biological properties by comparing the product with appropriate in-house reference standards could be implemented as a fundamental component of product development.

8.5 IMPACT OF PLANT-MADE BIOPHARMACEUTICALS ON HUMAN HEALTH

Plant-made biopharmaceuticals could potentially have an impact on human health. Several concerns are described here.

8.5.1 Transgenic Crops as a Potential Hazard to Diet

While plant-derived biopharmaceuticals are generally purified prior to administration, it is worthwhile to mention the potential for new hazards to appear in foods as a direct consequence of genetic engineering of crop plants. Risks associated with the appearance of toxins, allergens, or genetic hazards in foods derived from genetically engineered crops may arise as a consequence of the biosynthesis of specific chemical constituents in the portion of the crop that is eaten, or by the elimination of metabolites that are important in reducing various health risks (e.g., antioxidants) (Conner and Jacobs, 1999).

Three mechanisms by which insertion of transgenes may create food hazards are (1) the insertion of genes and their expression products (note:

the amount and stability of most protein products of transgenes can be determined and any potential hazard can be evaluated), (2) unexpected secondary and pleiotropic effects of gene expression. This may cause altered metabolic flow-through in biochemical pathways as a result of high levels of expression of transgenes expressing enzymes that catalyze biochemical reactions, and (3) disruption or modification of the expression of endogenous genes in the recipient plant or the activation of genes that are ordinarily silent, for example, by read-through from highly expressed promoter regions of transgenes into coding regions of flanking plant DNA via insertional mutagenesis. While these potential hazards are worth noting, it is important to keep in mind that random insertion events are not unique to genetic engineering, and do not present a health risk beyond those found from traditional plant breeding (Conner and Jacobs, 1999).

8.5.2 ALLERGENICITY

In the past, the FDA has approved drugs because their benefits outweighed their relative allergenicity potential. This general approach has also been taken with respect to plant-made biopharmaceuticals. Plant-derived biopharmaceuticals must be subjected to the same quality control and safety standards as materials derived from traditional bacterial or mammalian cell systems. The FDA will require a study to determine whether the identical drug produced in a plant will be more allergenic than its conventional pharmaceutical counterpart (e.g., differences in glycosylation patterns). To address these issues, the FDA has designed an allergenicity decision tree, which assists researchers in determining the potential risk to human health of the biopharmaceuticals they wish to generate in plants (refer to Figure 8.2 for more detail).

The question also arises as to whether there is a potential allergenicity risk for people who happen to inhale or topically receive pollen from genetically engineered plants (indirect contact) that may then cause an allergic reaction or immunity (Davies, 2005). Since the proteins produced in plants tend to be highly specific, the risk to human health is negligible (Goldstein and Thomas, 2004).

8.6 IMPACT OF PLANT-MADE BIOPHARMACEUTICALS ON THE ENVIRONMENT

There are a number of concerns regarding the impact of plant-made biopharmaceuticals on the environment. These are described in detail below.

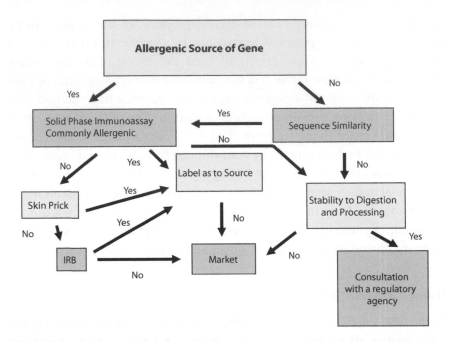

FIGURE 8.2 Example of an allergenicity decision tree. A series of questions resulting in "Yes" or "No" answers are posed to determine whether or not a plant-derived therapeutic product should or should not enter the marketplace.

8.6.1 GENE CONTAINMENT

Gene containment refers to procedures used to prevent the intermingling of food crops, nontarget organisms, and the environment with plant-based biopharmaceuticals. Confinement procedures are based upon scientific risk assessments that take into account the type of crop, the spatial setting or location of the production area, and agronomic and crop handling. Four major steps determine the likelihood of gene flow from crops to related species to prevent the production of hybrids and their progeny. These include (1) the presence of plants within pollen or seed dispersal range and the ability of actual gene flow by pollen or seed to take place, (2) synchrony or some degree of overlap in flowering time between crop and pollen recipient species, (3) sexual compatibility between crop and recipient species, and the ability to produce fertile hybrids, and (4) ecology of the recipient species to enable the establishment of crop genes in wild recipient populations (Dale et al., 2002; Gepts and Papa, 2003).

8.6.1.1 Spatial and Temporal Separation

This refers to the distance between the planting of crops that produce bio-pharmaceuticals and other, conventional crops for feed or the buffer zones between fields of the same species. When the use of physical isolation and distances between transgenic and nontransgenic crops cannot be realized, containment can take place in a greenhouse. Temporal separation refers to the time separating flowering and pollination between crops producing biopharmaceuticals and conventional crops of the same or related species. Transgenic crops can be planted at different times from food crops to ensure flowering occurs at different times; this decreases the potential for pollen transfer.

The uncontrolled spread of transgenic plants in the environment may take place by pollen transfer, by horizontal gene transfer to soil microorganisms, or due to selective advantages the transgenic plant may have over its wild relatives. These issues must be kept in mind when choosing geographical locations of test plots, plant species, and production protocols. Appropriate containment measures must also be taken for virus expression vectors to prevent infection of any susceptible nearby crops (Gepts and Papa, 2003). In addition, the persistence or invasiveness of crops must be addressed. For example, do newly introduced traits render a crop more likely to be more invasive in natural habitats? (Dale et al., 2002). It is important to note that if genes are transferred to a recipient plant, the recipient plant will only proliferate if the gene transferred confers a selective advantage. An event such as this should not take place with biopharming, and could even result in a selective disadvantage as the ability of the modified plant to proliferate and compete with its wild-type counterparts would be limited.

8.6.1.2   Prevention of Outcrossing

These standards can be met by adding male sterility traits to control pollen flow to avoid outcrossing. Some plant species that have enclosed flower structures would represent good candidates. A crop like corn, which is a wind-pollinated crop, can have tassels removed manually, and sterile varieties can be made to contain the pollen (Shama and Peterson, 2004).

8.6.1.3   Equipment and Training

Processing of plant-derived pharmaceuticals must take place under tightly controlled conditions, using production standards developed jointly by the USDA and the FDA. Included in this is the correct training of farm workers so that no commingling of transgenic crops with conventional crops takes place with respect to harvesting and farm equipment. The dedication of harvest and storage equipment solely to crops that produce biopharmaceuticals and controlled processing in a manner that keeps crops expressing biopharmaceuticals separate from other food crops, as well as extensive control of field access, harvest, and product disposition, is also required. In addition, containment strategies may include the methods of site preparation, planting, use of equipment, harvesting, product handling and distribution, and the appropriate handling of fields in years following the planting of pharmaceutical crops.

In the regulatory process, all classes of GM crop are examined on a case-by-case basis in the context of the geographical locations where they are to be grown. For example, the distance viable pollen can travel for each crop type is influenced by a pollen dispersal mechanism (wind, insects, etc.) and pollen longevity (determined by plant species).

8.6.2 NONTARGET ORGANISMS

Besides issues concerning the impact of pollen transfer with respect to the environment, the effect of the crop producing plant-derived biopharmaceuticals on nontarget organisms such as insects, and any adverse effect on birds and wildlife must be addressed. This includes the toxicity of the biopharmaceutical to living things, or the undesirable effect of a novel gene on "friendly" organisms in the environment (Dale et al., 2002). Animals that eat the plant or carry pollen from the plant (e.g., butterflies, honeybees) may be affected. In the past, this has been demonstrated to be an issue with genes conferring pest or disease resistance—for example, the effect of the Bt insecticidal protein on monarch butterfly populations (Shama and Peterson, 2004; Mascia and Flavell, 2004; Poortinga and Pidgeon, 2005). If the recombinant protein may present an environmental hazard, for example, a toxin produced in the plant, the appropriate environmental and agricultural authorities should be consulted during early product development. This risk can be reduced by planting nontransgenic border crops, or containing the plants under greenhouse conditions. Since the majority of plant-made biopharmaceuticals are species-specific (such as plant-made antibodies), the risk of a harmful effect is reduced. Some plant-derived vaccines may have side effects on the gastrointestinal tract, the result of antigenic specificity of the antibody (Ma et al., 2003). This still appears to be a remote possibility.

The potential of flow of plant-made biopharmaceuticals into the human food chain remains. For example, plant-derived pharmaceuticals could cross-contaminate foodstuffs by spontaneous growth of transgenic crops in areas outside the intended field, or by pollen flow between some plants such as corn. It has been suggested that plant-derived biopharmaceuticals should be generated in nonfood crops, such as tobacco. However, food crops produce the greatest opportunities for efficient production since they are among the most well-studied of crops. This continues to make them more feasible for edible vaccine production.

8.7 AVOIDING TRANSGENE SILENCING

Transgene silencing refers to the phenomenon by which transgene expression levels that are initially high are frequently impaired at a later stage of plant development or in later generations. Endogenous genes can also be silenced as a consequence of the presence of a transgene with homologous sequences (cosuppression). Differences in transgene expression can be due to the influence of the chromatin environment surrounding the transgene (position effect). Conversely, low transgene expression levels can be due to homology-dependent gene silencing (HDGS). Homology-based silencing mechanisms can act at the transcriptional or posttranscriptional level and can develop over subsequent generations (Meyer et al., 1998). In general, transcriptional gene silencing (TGS) involves the interaction of genes that share homology within the promoter region and is often associated with methylation patterns of these genes and their surrounding regions. Posttranscriptional gene silencing (PTGS) requires homology in transcribed regions of the silenced genes and involves enhanced, sequence-specific RNA turnover (De Wilde et al., 2000).

There are a number of steps that can be taken in the construction of plant-derived biopharmaceuticals in transgenic plants that would effectively reduce gene silencing (De Wilde et al., 2000). These are explained in the following sections.

8.7.1 IMPROVEMENTS IN THE DESIGN OF TRANSFORMATION VECTORS

When the production of multimeric proteins within the same transgenic plants is required, often all subunits must be produced at similar levels. In the past, use of the identical promoter and 3′ regions for each subunit encouraged gene silencing events to take place. To avoid gene silencing, such homologous regulatory regions cannot be included. Instead, tissue-specific or developmental promoters (e.g., seed-specific promoters) can be employed to avoid multiple use of the same promoter and, as a result, induce gene silencing.

8.7.2 EXAMINATION OF THE SITE OF INTEGRATION AND BASE COMPETITION OF THE TRANSGENE AND SURROUNDING REGIONS

It is important to note the methylation status of the site of integration. It has been demonstrated in the past that insertion of foreign genes into areas of plant genomes with characteristically high GC content may be a sign of an area of the genome that is inactivated by methylation. It is therefore paramount to ensure that the foreign DNA to be inserted into the host genome possesses a codon usage pattern that is optimized for the host species. If the transgene is inserted into a hypermethylated or condensed region of chromatin, it can also undergo transcriptional inactivation because the local chromatin would no longer be accessible for transcriptional factors. Solutions to this problem involve integration of the transgene through site-specific integration, or addition of matrix attachment regions at the site of integration.

8.7.3   Method of Transformation and Selection of Host Plant Species

Since different genotypes within a plant species have different gene silencing responses, various breeding and selection programs can be utilized to limit gene silencing in crops. In the future, mutant plants that are impaired in transgene silencing may be utilized for optimum expression of foreign proteins.

8.7.4   Growth Conditions of Transgenic Plants

It has been shown that plant cell cultures grown in bioreactors appear unable to spread a gene silencing response when compared to entire transgenic plants. This is most likely due to the fact that the signal used to elicit gene silencing is transmitted through the plasmodesmata and vascular tissue. Since cultured cells lack this form of interconnection, the signal for gene silencing cannot be transmitted.

8.7.5   Copy Number of Transgene

Studies have demonstrated that the presence of multiple copies of a particular transgene favor the gene silencing process. During plant transformation, multiple T-DNAs are often inserted in a repeated fashion into the genome at a single chromosomal locus. Employment of site-specific recombination in the integration procedure would resolve complex loci into single copy transgenes and reduce the likelihood of gene silencing (Garrick et al., 1998; Srivastava, 1999).

8.8   CONCLUSIONS

The ability to describe the risks involved in production of plant-made biopharmaceuticals quantitatively is important for comprehensive societal decision making and communication, and will help enhance public trust in the decision-making process surrounding the technology. However, fears still remain that gene flow from transgenic to nontransformed plants may somehow affect the food supply. For example, in 2002, a Nebraska farmer allegedly planted food soybeans in a field where biopharmed corn

(ProfiGene) had been grown the previous season but failed to provide adequate weed control to eliminate drug-bearing corn volunteers. Eventually, the corn-contaminated soybean was harvested and pooled with other soybeans. The entire batch had to be destroyed.

In another example, in 2004 a judge ordered the USDA to identify the Hawaiian locations of four companies operating open air test sites for biopharmaceutical crops. The order had been earlier denied as it was considered to contain confidential business information protected from disclosure under federal law. Public disclosure could result in the destruction of the fields by anti-GM extremists. Vandalism such as this does little to protect the health of the public or the environment. Rather, it causes the dispersal of transgenic crops into the environment, thus creating the very harm feared by these adversaries (Jaffe, 2004a).

Finally, a demand for on-the-plant detection systems in the field and identity tracking systems to monitor transgenes has prompted the development of a number of innovative technologies. For example, since gene expression can be affected by environmental conditions, it is of interest to measure transgene expression quantities and compare the real measurement with the expected value in the agricultural field.

A number of technologies for transgene detection have been developed. Some postharvest techniques such as PCR and ELISA can be easily adapted for living plant applications. However, these techniques require sampling and expensive laboratory testing. Many new technologies for transgene monitoring in living plants are based on optical or fluorescent markers (Miraglia et al., 2004). These and other technologies will make the cultivation of biopharmaceuticals in plants easier to track, thus providing another tier of control to keep gene flow to a minimum.

REFERENCES

Ciliberti, R. and Molinelli, A. (2005). Towards a GMO discipline: ethical remarks. *Veterinary Res. Commun.* 29(Suppl. 2): 27–30.

Conner, A.J. and Jacobs, J.M.E. (1999). Genetic engineering of crops as potential source of genetic hazard in the human diet. *Mutation Res.* 443: 223–234.

Dale, P.J., Clarke, B., and Fontes, E.M.G. (2002). Potential for the environmental impact of transgenic crops. *Nat. Biotechnol.* 20(6): 567-574.

Daniell, H. (2002). Molecular strategies for gene containment in transgenic crops. *Nat. Biotechnol.* 20: 581–586.

Davies, H.V. (2005). GM organisms and the EU regulatory environment: allergenicity as a risk component. *Proc. Nutr. Soc.* 64: 481–486.

De Wilde, C., Van Houdt, H., De Buck, S., Angenon, G., De Jaeger, G., and Depicker, A. (2000). Plants as bioreactors for protein production: avoiding the problem of transgene silencing. *Plant Mol. Biol.* 43: 347–358.

Docket 02D-0324 Drugs, Biologics and Medical Devices derived from bioengineered plants Comment Number: EC-43: volume 9.

Farrow, S. (2004). Using risk assessment, benefit-cost analysis, and real options to implement a precautionary principle. *Risk Anal.* 24(3): 727–738.

Garrick, D., Fiering, S. Martin, D.I., and Whitelaw, E. (1998). Repeat-induced gene silencing in mammals. *Nat. Genet.* 18(1): 56-59.

Gepts, P. and Papa, R. (2003). Possible effects of (trans)gene flow from crops on the genetic diversity from land races and wild relatives. *Environ. Biosafety Res.* 2: 89–103.

Goldstein, D.A. and Thomas, J.A. (2004). Biopharmaceuticals derived from genetically modified plants. *Q. J. Med.* 97: 705–716.

Howard, J.A. and Donnelly, K.C. (2004). A quantitative safety assessment model for transgenic protein products produced in agricultural crops. *J. Agric. Environ. Ethics* 17: 545–558.

Jaffe, G. (2004a). *Sowing Secrecy: The Biotech Industry, USDA, and America's Secret Pharm Belt*. Washington, DC: Center for Science in the Public Interest.

Jaffe, G. (2004b). Regulating transgenic crops: a comparative analysis of different regulatory processes. *Transgenic Res.* 13: 5–19.

Jones, P.B.C. (2004). Plant-made pharmaceuticals: progress and protests. ISB News Report. *Information Systems for Biotechnology*. October Issue.

Kirk, D.D., McIntosh, K., Walmsley, A.M., and Peterson, R.K.D. (2005). Risk analysis for plant-made vaccines. *Transgenic Res.* 14: 449–462.

Ma, J.K.-C., Drake, P.M.W., and Christou, P. (2003). The production of recombinant pharmaceutical proteins in plants. *Nat. Rev. Genet.* 4: 172–185.

Ma, J.K.-C., Chikwamba, R., Sparrow, P., Fischer, R., Mahoney, R., and Twyman, R.M. (2005). Plant-derived pharmaceuticals: the road forward. *Trends Plant Sci.* 10(12), 580–585.

Mascia, P.N. and Flavell, R.B. (2004). Safe and acceptable strategies for producing foreign molecules in plants. *Curr. Opin. Plant Biol.* 7: 189–195.

Miele, L. (1997). Plants as bioreactors for biopharmaceuticals: Regulatory considerations. *Trends Biotechnol.* 15: 45–50.

Miller, H.J. and Longtin, D. (2006). Down on the Biopharm. Policy Review Online

Miraglia, M., Berdal, K.G., Brera, C., Corbisier, P., Holst-Jensen, A., Kok, E.J., Marvin, H.J.P., Schimmel, H., Rentsch, J., van Rie, J.P.P.F., and Zagon, J. (2004). Detection and traceability of genetically modified organisms in the food production chain. *Food Chem. Toxicol.* 42: 1157–1180.

Peterson, R.K.D. (2000). Public perception of agricultural biotechnology and pesticides: recent understandings and implications for risk communication. *Am. Entomol.* 46, 8–16.

Peterson, R.K.D. and Arntzen, C.J. (2004). On risk and plant-based biopharmaceuticals. *Trends Biotechnol.* 22(2): 64–66.

Poortinga, W. and Pidgeon, N.F. (2005). Trust in risk regulation: cause or consequence of the acceptability of GM food? *Risk Anal.* 25(1): 199–209.

Rigano, M.M. and Walmsley, A.M. (2005). Expression systems and developments in plant-made vaccines. *Immunol. Cell Biol.* 83: 271–277.

Shama, L.M. and Peterson, R.K.D. (2004). The benefits and risks of producing pharmaceutical proteins in plants. *Risk Manage. Matters* 2(4): 28–33.

Srivastava, V., Anderson, O.D., and Ow, D.W. (1999). Single-copuy transgenic wheat generated through the resolution of complex integration patterns. PNAS 96(20): 11117-111121.

Wolt, J.D. and Peterson, R.K.D. (2000). Agricultural biotechnology and societal decision-making: the role of risk analysis. *AgBioForum* 3(1): 39–46.

Wu, F. (2004). Explaining public resistance to genetically modified corn: an analysis of the distribution of benefits and risks. *Risk Anal.* 24(3): 715–729.

9 Epilogue
The Future

9.1 THE CURRENT STATE OF PLANT-MADE BIOPHARMACEUTICALS

The development and commercialization of any new drug or therapeutic agent, even one designed by conventional means, is a slow and tedious process. With the inclusion of FDA (U.S. Food and Drug Administration) approval, this course of action can take between 10 and 15 years for the full development of a new product to be complete. In addition to this, the cost for the development of a new pharmaceutical product for commercialization can be as great as $1 billion. Moreover, only a small percentage of new pharmaceutical products that undergo human clinical trials are able to successfully advance to the commercialization stage. A significant number of candidate vaccines that fail to reach the market exhibit problems with respect to efficacy or safety of the product during the clinical trial process.

Plant-made biopharmaceuticals have been under research and development for close to 20 years, and the first plant-derived vaccines for human or animal use have only now become available. Much of the delay in the commercialization of any vaccine is dependent upon the integrity of the regulatory process. In general, vaccines that are deemed for veterinary use follow a shorter regulatory process and receive regulatory approval long before vaccines that are developed for human use. It comes as no surprise, then, that one of the first vaccines derived from plants to be commercialized is a Newcastle disease virus vaccine, targeted to protect poultry. The first plant-derived vaccines, then, are now leaving the realm of academia and making their initial appearances in the corporate world.

One of the reasons that the development of plant-derived biopharmaceuticals has been prolonged is the requirement for a substantial multidisciplinary research approach. It is improbable that any single scientist has the expertise to bring the research from the initial laboratory bench through to the final stages of completion, resulting in a "prototype" product that could then enter the commercialization process. Research and development of plant-derived biopharmaceuticals include input from fields as far-ranging as plant science, microbiology, glycobiology, immunology, toxicology, crop science, pharmacology, and engineering. It is in fact becoming increasingly important that researchers have a complete plan from the start of exactly how they intend to take their work (or pass it along) through to later stages such as Phase II clinical trials. The requirement for multidisciplinary cooperation will continue to evolve, as further research will bring forth new questions regarding the safety, efficacy, and allergenicity of new plant-made vaccines and therapeutic agents. For example, further and more complete understanding of the humoral and cell-mediated immune responses to plant-derived vaccines will require constant input from both immunologists and pharmacologists as their respective fields advance.

The cooperation of scientists from many disciplines is paramount to ensuring that a potential plant-derived biopharmaceutical product continues along the path to commercialization. However, many challenges remain to be addressed by those who prefer to adhere to more of an academic research perspective. For example, more research regarding the mechanisms of targeting, protein folding, assembly, and storage of the desired biopharmaceutical product in the host plant are required. Procedures can be improved to produce optimum yields of vaccine and therapeutic proteins in crops, as well as more efficient harvesting, extraction, and downstream processing from plant tissue. A selection of new production hosts and virus vector expression systems will without doubt become available as these fields expand and develop. Our current modest understanding of mucosal immunity will improve, as will our knowledge of allergenicity and oral tolerance as researchers shed light on these areas. Insight into each of these fields will greatly assist in the general process of using plants as production platforms for biopharmaceuticals.

As mentioned throughout this book, a number of hurdles still remain that may impede the progress of plants as biofactories for production of vaccines and therapeutic proteins. For example, the intellectual property environment for plant-made biopharmaceuticals can be complex, making it rather cumbersome and difficult to progress in the commercialization process. Furthermore, a negative opinion over genetically modified plants continues to dwell in some portions of the public sector. It is clear that private industry is more interested in providing products that will successfully enter the commercial market in the industrialized world, and as a result are vulnerable to negative public opinion. However, the intense need by the poor and those in developing countries for vaccines and therapeutic agents that are inexpensive and easy to mass produce makes it unlikely that plant-derived biopharmaceuticals will become a "flash in the pan" technology. This potential application in humanitarian aid may in fact provide the use of plants as protein production platforms the necessary boost to gain distinction in the world.

Index

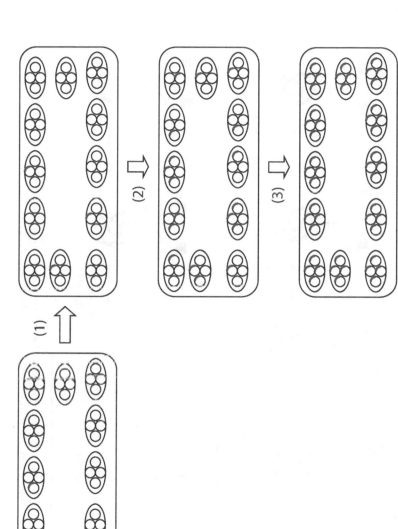

FIGURE 3.1 Establishment of a transformed, homoplasmic cell line. The primary chloroplast transformation event involves the change of a single copy of the plastid genome out of thousands of copies in a single leaf cell. Stages (1)–(3) each represent cell divisions under selective antibody pressure. After each cell and organelle division, the selective antibiotic favors the multiplication of those chloroplasts that contain the transformed copies of the genome. At stage (2), some chloroplasts may contain a mixed population of transformed and wild-type genomes (heteroplasmy). At stage (3), those chloroplasts that harbor wild-type genomes are eventually eliminated, and homoplasmy is achieved over several rounds of plant regeneration on selective medium.

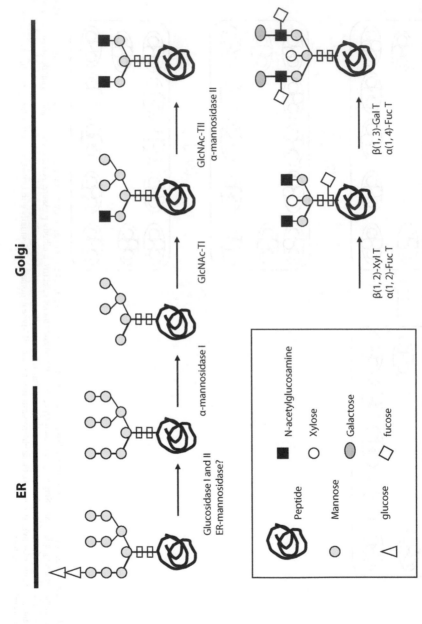

FIGURE 5.1(b) Biosynthetic processing of *N*-glycans in the plant ER and Golgi. Each step indicates the glycosyl groups added to or modified on a nascent polypeptide and the enzymes responsible for this process. (Revised from Chen et al. (2005). *Med. Res. Rev.* 25(3), 343–360.)

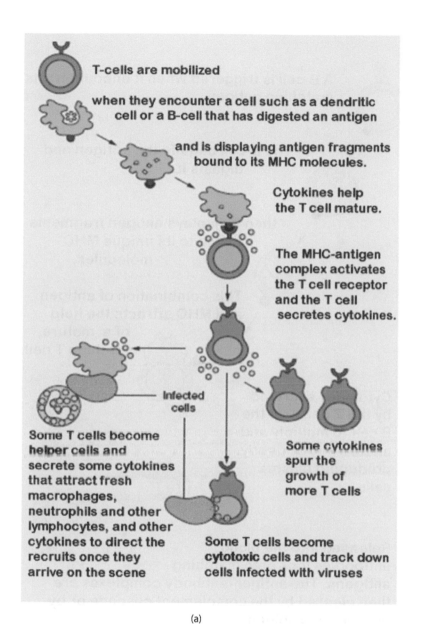

T-cells are mobilized

when they encounter a cell such as a dendritic cell or a B-cell that has digested an antigen

and is displaying antigen fragments bound to its MHC molecules.

Cytokines help the T cell mature.

The MHC-antigen complex activates the T cell receptor and the T cell secretes cytokines.

Infected cells

Some T cells become helper cells and secrete some cytokines that attract fresh macrophages, neutrophils and other lymphocytes, and other cytokines to direct the recruits once they arrive on the scene

Some cytokines spur the growth of more T cells

Some T cells become cytotoxic cells and track down cells infected with viruses

(a)

FIGURE 7.1(a) Development of cytotoxic and helper T cells. T cell maturation takes place upon activation of the T cell by an antigen-presenting cell such as a dendritic cell or a B cell. Antigen fragments are bound to the MHC complex and displayed on the cell surface of an antigen-presenting cell. This MHC-antigen complex interacts with the T cell receptor of the T cell and stimulates it to release cytokines. Cytokines secreted by the activated T cell induce the proliferation of both cytotoxic T cells, which kill infected cells, as well as helper T cells, which direct other immune cells at the scene of infection.

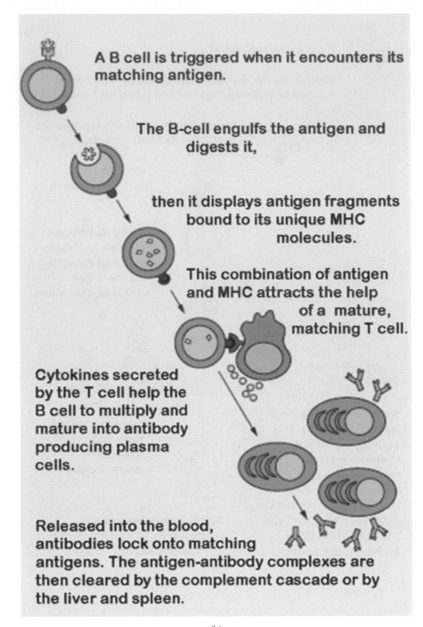

A B cell is triggered when it encounters its matching antigen.

The B-cell engulfs the antigen and digests it,

then it displays antigen fragments bound to its unique MHC molecules.

This combination of antigen and MHC attracts the help of a mature, matching T cell.

Cytokines secreted by the T cell help the B cell to multiply and mature into antibody producing plasma cells.

Released into the blood, antibodies lock onto matching antigens. The antigen-antibody complexes are then cleared by the complement cascade or by the liver and spleen.

(b)

FIGURE 7.1(b) B cell activation and the generation of antibodies. Upon encountering an antigen, a B cell ingests the antigen and displays fragments of it upon its surface via a MHC complex. A T cell that interacts with the antigen-MHC complex of this B cell then releases cytokines to direct the B cell to proliferate and mature into antibody-producing plasma cells.